*David*
**WALTNER-TOEWS**

# E A T

*—— the ——*

## BEETLES!

*David*
**WALTNER-TOEWS**

# E A T

## *the*

## BEETLES!

*An* EXPLORATION *into*
*Our* CONFLICTED RELATIONSHIP
*with* INSECTS

Published by ECW Press
665 Gerrard Street East
Toronto, Ontario, Canada, M4M 1Y2
416-694-3348 / info@ecwpress.com

Get the
eBook free!*
*proof of purchase
required

Purchase the print edition and receive the eBook free!
For details, go to ecwpress.com/eBook.

LIBRARY AND ARCHIVES CANADA
CATALOGUING IN PUBLICATION

Waltner-Toews, David, 1948–, author
Eat the beetles! : an exploration into our conflicted
relationship with insects / David Waltner-Toews.

Issued in print and electronic formats.
ISBN 978-1-77041-314-6 (paperback)
Also issued as: 978-1-77305-036-2 (pdf)
978-1-77305-035-5 (epub)

1. Entomophagy. 2. Edible insects. 3. Beneficial
insects. 4. Insects—Ecology. 5. Insects—Effect
of human beings on. I. Title.

GN409.5.W35 2017    641.3'06
C2016-906411-5    C2016-906412-3

Cover design: David A. Gee
Cover images: empty tray © Oktay Ortakcioglu /
iStockPhoto; gripping © blackred / iStockPhoto;
cloche © urfinguss / iStockPhoto
Interior images: frame © funkyboy2014/Vecteezy.com
Author photo: Kathy Waltner-Toews

The publication of *Eat the Beetles!* has been generously supported by the Canada Council for the Arts, which
last year invested $153 million to bring the arts to Canadians throughout the country, and by the Government
of Canada through the Canada Book Fund. *Nous remercions le Conseil des arts du Canada de son soutien. L'an
dernier, le Conseil a investi 153 millions de dollars pour mettre de l'art dans la vie des Canadiennes et des
Canadiens de tout ce pays. Ce livre est financé en partie par le gouvernement du Canada.* We also acknowledge
the support of the Ontario Arts Council (OAC), an agency of the Government of Ontario, which last year funded
1,737 individual artists and 1,095 organizations in 223 communities across Ontario for a total of $52.1 million,
and the contribution of the Government of Ontario through the Ontario Book Publishing Tax Credit and the
Ontario Media Development Corporation.

Ontario
Ontario Media Development
Corporation

ONTARIO ARTS COUNCIL
CONSEIL DES ARTS DE L'ONTARIO
an Ontario government agency
un organisme du gouvernement de l'Ontario

Canada Council
for the Arts

Conseil des Arts
du Canada

Canadä

PRINTED AND BOUND IN CANADA

PRINTING: MARQUIS   5   4   3   2   1

RECYCLED
Paper made from
recycled material
FSC® C103567
www.fsc.org

# IT'S FOR YOU

This book is dedicated to my grandchildren,
Ira, Annabel, Wendell, and Nikolas.

# Table of Contents

CRICKET TO RIDE

*An Introduction*

· · · · · · ·

*Have you seen the little chiggers?*

My legs were aching and my head was still dazed from the trans-oceanic Toronto-to-Paris flight. Like many before me, I had come to the City of Light for the cuisine. Unlike many others, I came in search of grubs.

I walked along boulevard des Batignolles and then up the slopes of rue Caulaincourt, a curved street on Montmartre, near the Sacré-Cœur Basilica and its attendant nightclub district, appropriately situated near a place of confession. After asking directions at several shops in my Canadian franglais, I was waved farther up the hill and then down a flight of stairs, across rue Lamarck, then down another set of stairs, across rue Darwin, and down the cobbled slope of rue de la Fontaine-du-But. If there was some significance to my passing the ghosts of two greats of nineteenth-century evolutionary biology, I missed it. I was busy looking for a pub named after William S. Burroughs's hallucinatory novel *Naked Lunch*.

Le Festin Nu had been described by *Business Insider* in 2013 as "a trendy establishment in the 18th arrondissement of Paris," and the first restaurant in Paris to include insects on the menu. About the same time, the restaurant was featured on the website Fine Dining Lovers, which described eating insects as France's "rising trend" in fine dining, and in a BBC report on the insects that were appearing on the menus of "upscale French restaurants." Still, weird food fads and drug-induced novels notwithstanding, why would a normal-looking, sort of sane guy fly to Paris to eat bugs in a pub?

In 2013 the Food and Agriculture Organization of the United Nations (FAO) published *Edible Insects: Future Prospects for Food and Feed Security*. According to that UN report, "insects form part of the traditional diets of at least 2 billion people. More than 1,900 species have reportedly been used as food."[1] I didn't pay much attention. Yes, some people eat insects, I thought. In a world roiling with multiple ecological and political catastrophes, why should I care about this culinary curiosity? Then, in March 2014, Veterinarians without Borders/Vétérinaires sans Frontières-Canada (VWB/VSF), under a larger program of "Improving Livelihoods and Food Security in Laos and Cambodia," initiated a project on small-scale cricket farming. I was skeptical; were crickets sufficiently "animal" to qualify as subjects of veterinary interest? Weren't insects mostly pests and disease carriers? When I stepped out of my comfort zone to investigate, I was swept away in a chaotic deluge of reports, blogs, videos, books, and papers: 2014, it appeared, was a global tipping point for entomophagy, which is what enthusiasts were now calling insect-eating.

In May of 2014, the lead author of the FAO report, Arnold van Huis, organized the first world conference on whether insects could feed the burgeoning world population. Across the Atlantic in America, Daniella Martin, bug-eating enthusiast, blogger, and host of an insect/travel show called *Girl Meets Bug*, published *Edible: An Adventure into the World of Eating Insects and the Last Great Hope to Save the Planet*. On her website, Martin announced that she had eaten "bees, crickets, cockroaches, fly pupae, wax worms, mealworms, silkworms, hornworms, bamboo worms, grasshoppers, walking sticks, katydids, scorpions, tailless whip scorpion, snails, stink bugs, tarantulas, cicadas, leaf-cutter ants, ant pupae, dung beetles, termites, wasps and wasp brood,[2] butterfly caterpillars, dragonflies, and water bugs." *Wow*, I thought, my hand pressed to my heart. *Really? What did she* not *eat?*

A tornado of amazing claims swept across the usually staid landscapes of agriculture and food. Insects are higher in protein than other livestock, I was informed, contain more unsaturated (good) fats, and often are rich in "blood-building" minerals such as iron and zinc. Their production is associated with lower greenhouse gas emissions, and they need less land and less water, and in general use fewer resources — some even thriving on food waste — than other livestock. They require minimal physical infrastructure to grow, and provide both increased income and improved nutrition for subsistence-farming households in poor parts of the world. Women in rural areas of the tropics, announced a writer in that venerable magazine *The Economist*, could reap particular benefits, related to both personal health (from increased consumption of iron and calcium) and economic well-being.[3]

Furthermore, I discovered, this was not just another "let's help poor people" boondoggle. In May of 2015, in Langley, British Columbia, I watched as trucks dumped vegetable waste at Enterra Farms, where it was fed to fly larvae, thus turning waste into protein concentrates for farmed salmon and chickens — displacing the conventionally used fishmeal and soybean powder, both of which are associated with serious negative environmental impacts.[4] In the rolling countryside outside Peterborough, Ontario, Entomo Farms[5] had transformed a chicken barn into an environmentally conscious family-owned farm producing roasted crickets and mealworms, as well as protein powder (ground-up crickets) for human consumption. Standing next to the stainless steel bake ovens, I sampled their roasted, Moroccan-spiced crickets.

At the other end of the technological spectrum, in France, was Antoine Hubert, winner of a 2015 *MIT Technology Review* Innovators Under 35 award, president of the European Association of Insect Producers, and CEO of a company called Ynsect. Hubert was championing biotechnological approaches to using insects as sources of quality bioactive and nutritional products; in his view, this high-tech use of insects is a "disruptive technology" that will transform global agriculture. Coined by Harvard Business School professor Clayton M. Christensen, the term *disruptive technology* is defined as a "technology that can fundamentally change not only established technologies but also the rules and business models of a given market, and often business and society overall."[6]

Where was all this leading?

In an interview that serves as part of the introduction to *The Insect Cookbook: Food for a Sustainable Planet*, former Secretary-General of the United Nations Kofi Annan envisions a world in which insects as food are traded globally, providing economically important and environmentally sustainable nutrition and food security for communities around the world.

So, even before I'd sampled the bar snacks at le Festin Nu, my head was in turmoil. All around me, bugs were changing everything I thought I knew about food, feed, and agriculture. How could I have missed this?

The Festin Nu bar was a room just a few meters wide, clad in dark, weathered wood with handwritten notices taped on the plate glass windows framing a darkened doorway. On a couple of low stools out on the sidewalk, a man and a woman were leaning against the glass, sipping beer, and chatting quietly in the warm, late afternoon haze of the day. The stores on either side had metal shutters pulled down. Inside, a reverse-J-shaped bar took up the far end of the small room. Beyond that, through a dimly lit doorway, a dozen or so people were sitting on church basement chairs, drinking beer and watching *Romancing the Stone* with French subtitles: how young Michael Douglas and Kathleen Turner looked! (Danny DeVito never changes.)

The bartender, Alex Cabrol, with his wry almost-smile, dark, wavy hair, and ragged straw fedora with a hole in it, reminded me of a 1967-vintage "Motorcycle Song" Arlo Guthrie. When I asked for insects, he waved at the menu board and asked which ones.

"All of them," I said, hoping I did not sound too much like a bug-eating version of Monty Python's Mr. Creosote. "We could go two ways with insect-eating," le Festin Nu's chef, Elie Daviron, said in that BBC report. "The agro-industry would just churn them into protein flour. I want to keep the notion that the insects are real, whole animals." Real, whole animals. I tried to imagine them and then decided that would be a bad strategy. *Brace yourself, order, and eat*, I told myself. *It's just food.*

August being vacation season for many Parisians, Cabrol explained, Daviron was unavailable, and the kitchen was not fully stocked. Of the six insects listed on the menu, only five were available. They didn't have any of the giant water bugs left in stock. He poured me one of the craft beers for which they are known, and I sipped it, wondering if I would be able to convince my brain that eating worms and crickets was okay.

I asked Alex where the bugs came from. He said that some came from Southeast Asia — he was leaving the next day for Cambodia to check out new possibilities — but they also sourced crickets and mealworms from a French breeder and distributor called Dimini Cricket. My first thought was that the name was a too-cute reference to Jiminy Cricket, the talking insect in Disney's *Pinocchio*. I recalled that in Carlo Collodi's original story, the young, wooden-brained vandal, in a fit of annoyance, had killed the cricket by throwing a mallet at him. In Disney's version, the cricket was a funny, wise, tag-along, apparently insufficiently annoying to incite murder. But would you eat him? In Disney's moral universe, I wondered, would it be as bad to eat crickets as to eat Bambi? I later discovered that the name of the breeder was based on the surname of one of the farm's founders, which is a cautionary tale about the perils of cross-cultural branding.

After about twenty minutes, Alex brought out five small plates, presented like tapas. The insects were artfully arranged, each species accompanied by figs, sun-dried tomatoes, raisins, and chopped, dried tropical fruits. I quaffed my beer and considered the fare before me: buffalo worms, crickets, large grasshoppers, small black ants, and fat grubs with beaks, which I later identified as palm weevil larvae.

I called for another beer, and then, bite by mindful, methodical bite, I ate them all. The crickets and grasshoppers were crunchy, with no strong flavor, the ants sour, tangy. The palm weevil larvae were a bit chewy, like dried figs. In the manner of pub food the world around, the dishes were on the greasy side. I guessed that with a couple of pitchers of craft beer, and a group of friends, these bugs would be just fine. But were they the future of global food security?

Later, strolling the packed alleys and streets among the tourists in and around Montmartre, I was chuffed that I'd passed my Paris pub bug challenge without having suffered any hallucinatory visions of giant cockroaches. Ever since I'd heard the name of this pub, I couldn't get out of my head images of those human-sized insects from David Cronenberg's adaptation of *Naked Lunch*. I wondered why I thought of eating insects as strange, even revolting. What made eating bugs in a pub any different than eating deep-fried chicken wings? Was this mix of beer, dare, and disgust the future of eating insects? Or would eating insects morph from the queasy adventures of "bug-eating" into the more neutral, sanitized cuisine of *entomophagy*, the term used for insect-eating by many of its proponents?

Less than a week after my visit to le Festin Nu, I was in Vientiane, capital of the Lao People's Democratic Republic, having a relaxing evening with Thomas Weigel, manager of the VWB/VSF cricket-farming project. We sipped mugs of Beer

Lao and shared a large plate of crickets, fried up with garlic and kaffir lime. Sitting just inside the pub, I watched the Laotians at the tables sprawled across the grooves of the wide sliding doors and onto a concrete patio toward the street. Thai singers belted out unstable melodies from a television set above the cashier's counter, the words dancing across the screen. Several groups of people in their twenties, happily half-drunk, were singing lustily along, wildly off key, about heartbreak and lost love. No one thought it strange that Thomas and I were noshing on fried crickets. This was not a challenge; there was no disgust to hurdle over. In Vientiane, this was normal. But on a world scale?

During my research, which over the course of 2015–16 included foraging in Japan; visits to insect farms, pubs, and restaurants in British Columbia, Ontario, France, England, Laos, Japan, and Australia; and reading many hundreds of books, scientific papers, and news reports, I was beginning to wonder if we were entering a "new normal" — and, if so, what that would look like. Would there be — as Kofi Annan imagined — a fairtrade network linking family farmers and foragers in North America, Europe, Southeast Asia, and Africa? Would some of us opportunistically gorge on locusts, or cicadas, while others foraged seasonally on termites, hornets, bees, or grasshoppers? Would I be able to walk into the local grocery store and, next to the 100 varieties of potato and corn chips, find bags of barbecue-flavored crickets? Were there insects that were off-limits? If so, then why? How would insects in the food system be managed? Loosely, like honey bees,[7] or intensively, like silkworms and crickets? Would most insects be ground up and used as energy and protein ingredients in animal feeds and energy bars for fitness enthusiasts?

As I discovered on my journeys, the world of insects in food and feed was more complex, exciting, and unsettling than a garnish of bugs on a dinner plate. To paraphrase a saying by renowned biologist J.B.S. Haldane, insect-eating is not only more complex and strange than we imagine; it is more complex and strange than we *can* imagine.[8]

In his farewell address as Professor of Tropical Entomology at Wageningen University in the Netherlands, Arnold van Huis concluded that the "benefits of using insects as food and feed over conventional meat are numerous. Insects have high potential of becoming a new sector in agriculture and the food and feed industry."[9] Listening to van Huis and his colleagues, one might be led to believe that the biggest challenges to humans eating insects are a combination of intransigently conservative European and North American consumers and outdated or missing regulatory frameworks and trade agreements. These are, indeed, important considerations, but a more careful reading of the situation should give us reason to reconsider this framing of the story. The executive summary of that 2013 FAO report asserts that "the huge potential" of entomophagy cannot be achieved without more research and better documentation of the nutritional value of insects as well as their environmental impacts relative to other sources of protein. Van Huis has noted that "clarification and augmentation of the socio-economic benefits that insect gathering and farming can offer is needed, in particular to enhance the food security of the poorest of society."[10]

For those who have grown up in non-insect-eating cultures, much of the entomophagical energy in the past few decades has focused on the fork end of the farm-to-fork food chain. Repeatedly, advocates return to the question: why don't people eat insects?

By which they mean: why don't people of European descent eat insects? Digging up and destroying the cultural roots of disgust about bugs as food and persuading people to make insects part of their daily meal plan becomes, in this perspective, a public relations exercise — one that could play out as a mix of research into consumer attitudes and glossy promotional advertisements, some of them bordering on a sort of moral blackmail, implying a lack of ecological caring among those who don't eat insects.

Another side of this complex problem is what has been called supply-side sustainability: how much food can the planet produce in a sustainable fashion? This raises different questions. Are the insects intended for human food or for use as animal feeds? Are they foraged or farmed? If foraged in the wild, does human foraging put at risk the functions of these insects in natural systems? Most of the eco-arguments, particularly in Europe and North America, rest on the assumption that we will be farming them. Will these farms be computer-regulated high-tech laboratories, such as Ynsect is developing in France, or more conventional family farms, like Entomo in Canada? As we have discovered over the past century, not all ways of farming are equally eco-friendly and sustainable.

To put these questions into context, some reflection on other sorts of livestock farming might be useful.

We inherited from our ancestors the idea that cows and chickens and pigs were the proper animals to provide us with food; these animals were the neighbors down the street, the families we knew. Henry IV of France declared, in about 1600, that he aspired to put a chicken in the pot of every peasant every Sunday. Since then, European and North American chicken (and pig, and cow) rearing practices have focused single-mindedly on the goal of getting low-cost meat into every pot, preferably every day. At first glance, this goal seems not so different from the entomophagists' dream of enhancing global food security. But a hundred years ago, when we were designing our farming and food systems, we didn't understand much about ecological and social complexity, energy and nutrient feedback loops, and the unintended consequences of focusing on one goal at a time. In retrospect, our way of getting a chicken, cow, or pig into every pot has been reckless, without any consideration of the cost in environmental destruction, climate change, disease, and inequalities based on money, gender, ethnic background, and political power. Our agri-food systems have strengthened traditional patriarchal power — and this is more than a metaphor — using science and technology not as ways to open our minds and learn new things, but as ways to secure the walls of the agro-industrial fort. It's not called "biosecurity" for nothing.

The way we have organized our cattle rearing has led to massive shifts of water and nutrients out of some ecosystems, and created great piles of water-contaminating dung in others. The toxin-producing strains of *E. coli* that cause severe diarrhea and kidney disease were first identified in hamburger but now turn up everywhere in our agri-food system. Toxic nitrogen leaks into waterways and down into aquifers, so that in many places where intensive agriculture is practiced, potable water is scarce. The way we have raised chickens has led to worldwide pandemics of food-borne bacteria such as *Salmonella* and *Campylobacter*. In the United Kingdom, levels of *Campylobacter* contamination of poultry carcasses have been so high that in 2015 the UK Food Standards Agency recommended consumers not wash raw chicken, but cook it thoroughly in its unwashed state. The concern was that washing would only spread the bacteria around the kitchen. And then of course there are avian influenza (bird flu) and seasonal influenza pandemics, which affect hundreds of thousands of people annually, and which are the result of viruses migrating from waterfowl to pigs, chickens, and people. These are unintended consequences of well-intentioned systemic changes we have made in how we grow and distribute food.

Until recently few questioned the choices of our evolutionary and historical forebears. What happened, happened. What is, is. Now, for the first time ever, we can make a deliberate, informed choice about which, if any, six-legged mini-livestock (another name sometimes used for farmed insects) we want on our dinner plates. As we consider putting bugs on the menu, we have an unprecedented opportunity to learn from our history. We can choose whether we want to hunt them or farm them, and if we do farm them, we can choose how, and where. Unlike our forebears, who stumbled ad hoc into a world-changing agricultural revolution, we have in hand the benefit of a century of intense scientific, economic, and cultural investigation. Not all farms have to be big. Not all foods have to be globally distributed, in abundance, or universally loved.

Eating is, in a sense, the human–environment equivalent of sex. The environment — in the form of animal, vegetable, and mineral foods — slips into our bodies and reconfigures itself as our flesh and blood. What we eat becomes who we are. So before we skip eagerly into the kitchen and get cozy, we should get to know these bugs a little better. Who are they? What were they doing before they showed up at our door? Did they travel far? Were they raised humanely? If our immediate response is disgust, why is that? Is it a deep evolutionary reaction to possibly poisonous foods, a by-product of scarcity in certain ecosystems, or maybe only a way to demarcate the insiders and outsiders of social groups? And if it *is* learned, is it because industrial agricultural CEOs want to protect their power and money, to prevent competition, or are there deeper ethical issues at play?

Some of the questions surrounding entomophagy are technical and scientific. Some are cultural, ethical, and what some would call spiritual. Still others — perhaps the most challenging ones — are organizational, legal, and administrative. In the early 1980s, I worked with 100 dairy farmers across Ontario. Some of them wanted to market their milk as organic. Consumers, they said, wanted it, and the farmers could produce high volumes of milk using organic methods, but, just as governments today seem unsure of how to regulate insects as food and feed, no one was quite sure how to fit organic products into an already-established structure designed for large-scale standardization rather than diversity. The farmers needed a legally recognized certification process. They needed processing plants dedicated to organic milk producers. If grocery stores were going to sell the products, they wanted a steady supply. Was this going to be just another case of "get big or get out?" Eventually the Ontario producers solved the economy-of-scale problem by creating a cooperative of family farms, working with other dairy farmers and governments to develop a regulatory framework and with processors to create mechanisms to get their products from the cows to the consumers. Now I can walk into any grocery store in the province and buy their milk and cheese as I would any other food. Would something similar happen with insect producers as they moved from the ambiguous margins to the mainstream of our food system?

In this book I shall raise all of these questions, and many more. As we ponder the crickets and mealworms on our plates, I hope that we see them as something more than a sustainable source of protein. I hope that we are also shocked, dismayed, and amazed; that we question who we are, and what entitlements and biological and social contracts we are heirs to. I hope that the bugs on our plates help us feel both uncomfortable and more at home.

If Daniella Martin's book *Edible* is your speed-dating guide to entomophagy, think of this book as Dad's cautionary dating manual. Think of me as Mr. Applegate from that '90s movie *Meet the Applegates*: the praying mantis pretending to be a normal suburban dad, waiting at the door, asking, "So, who is this guy you want to marry?"

With entomophagy, we have a chance to ask questions, to do better things rather than just doing the same old stuff more efficiently. It would be a shame to blow that chance.

## PART I. MEET THE BEETLES!

So, we are going to eat insects, are we? Which ones? What are their names? How many are there, and why are there so many of them? Are they really as good for us to eat, nutritionally, as some people say? Is entomophagy good for the planet? Is eating insects our last, best flight out of dystopia to the paradise of Ob-La-Di, Ob-La-Da? Let us begin our exploration.

*I Call Your Name*

. . . . . . .

*For if I ever saw you /*
*I didn't catch your name*

Alan Yen and I were enjoying a pleasant summer-evening drink on a friend's deck in Melbourne, Australia. Alan is an Australian biologist who has spent decades studying insects and human–insect relationships. When I told him about the book I was writing, he asked me why I had titled it *Eat the Beetles*. Was I only going to write about edible beetles? What about all those other insects?

Indeed, why beetles? It began, I confess, as a bit of a lark, all those puns involving the Beatles being too much fun for me to resist. It was good marketing. In my own mind, I could even come up with a convoluted rationalization. The Beatles were the Cambrian explosion of popular music. From a rocky start in a dark place, their music evolved over just a few years, invoking Lady Madonna and Ukrainian girls, liberation theology and humanism and atheism, Catholicism, Hinduism, communism, and entrepreneurialism. They brought us a mix of rock, blues, folk, classical, plugged and unplugged; movies that were like a collage of rock videos well before rock videos existed; big orchestras, intimate quintets; electronic music, piano, strings; biting, stinging, sweet, and sentimental; and all the other musical organisms one might expect from an enthusiastic band of Coleoptera. So this book would be a paperback version of the 1982 movie *The Compleat Beatles*, a tale of kicking back and feeling at home, perhaps while indulging oneself on insect snacks during a weekly celebration of Entomo-Tuesday. (Sir Paul, are you listening?)

All that was post-hoc rationalization, however, and this is a science book. How could I justify the title not just in punny marketing terms, but as a shorthand way to describe the explorations in this text?

The 1,900 known-to-be-edible insect species documented in the 2013 FAO report are but the crust on the *crème brûlée* of possible entomophagical options. Given the multitudinous diversity of insects out there, the possibilities for culinary experimentation would seem to be almost infinite. Which raises a number of questions: How many different species of insects are there? What are their names? And what is the total population of insects of all species on this planet?

If we are going to eat them, we should know who they are, where they live, and what they do for a living when not being eaten. Not to "name names" when looking at entomophagical options would be like saying we can eat mammals, which would include rhinos, pandas, tigers, orangutans, dogs, cats, mice, human babies, and monkeys, as well as cows, sheep, and pigs. Of course we *can*, but we also have important reasons for *not* eating some of them that have little to do with their nutrient content or our food preferences. The same is true for insects and, as we shall see, this has important implications for the future of ento-mophagy.

The names people attach to the world around them reflect the way they see the world. Economists divide the world into "haves" and "have-nots." During the Cold War, the world was divided politically into the First World (Europe, United States, and their allies), the Second World (USSR, China, and their allies) and the Third World (non-aligned countries, often from the global South). Another way to divide up the world is into those cultures that have a tradition of eating insects (the insect-eaters) and those that don't (the non-insect-eaters). This way of dividing up the world does not always coincide with polit-ical and economic boundaries, but is useful for understanding some of the major challenges facing twenty-first-century ento-mophagy promoters. In general, most insect-eaters live in trop-ical or subtropical parts of the world, such as Southeast Asia and sub-Saharan Africa, and most non-insect-eaters are from temperate zones, such as Europe, Russia, and northern parts of North America.

How indigenous people, urban consumers, farmers, and sci-entists classify natural entities is a reflection of how each group recognizes and relates to the world around them. Insects may be grouped as pests, food, or medicine, for instance, with sub-groups within those. These classifications are not inherent in the bugs, but in the roles we see them playing in our lives. The details — which are where, we are told, the devil lives — are relevant to defining an entomophagy that is both culturally and ecologically resilient. The details are also overwhelming (at least to me), but that may be because, unlike Mick Jagger and John Milton, I have never had much sympathy for the devil.

One might suggest that insect-eaters are the experts in identifying and classifying edible species, or stages of species (larvae versus adults, for instance), and appropriately preparing insect dishes from potentially toxic species. Insect-eaters' knowledge and classification systems do bring with them important information. But they also bring their own problems.

Some insects poke and suck, some bite and chew, some fly, some only crawl or hop. Some species undergo incomplete metamorphosis, where the babies look like little adults (think of crickets), while others, like butterflies and moths, undergo complete metamorphosis, where the baby caterpillars grow up to look like a different species. Thus, in some parts of the world, people may be able to identify an edible grub by name, but not its adult counterpart. In southern Africa, the fat, sausage-like larvae of emperor moths (*Gonimbrasia belina*), called mopane caterpillars or mopane worms because they feed on mopane trees (*Colophospermum mopane*), look nothing at all like the adult moths. Alan Yen has documented that Australian Aboriginal languages sometimes have different names for the same species, and that these names are not consistently translated into English the same way. Non-insect-eater scientists recognize one species of witjuti grub (*Endoxyla leucomochla*), the larva of a cossid moth that lives in the roots of *Acacia kempeana* (or the witjuti bush). At least two Aboriginal groups use a different binomial name for grubs, the first name referring to the fact that it is edible, and the second name being the plant on which the grub usually feeds. Conversely,

Yen writes, "indigenous people in central Australia recognize at least 24 different types of edible caterpillars from plants, and it is likely that most of these will be distinct scientific species."[11]

By contrast, non-insect-eaters tend not to identify insects according to whether or not they are edible. Mealworms, for instance, are so named not because someone thought they made a good meal, but because they have a long history of living in — and eating — the ground seeds and grains (a.k.a. meal) that people wish to keep for themselves.

For non-insect-eaters, differences in naming reflect both scientific rules and different subcultures within the scientific community. Thus, the terms *population*, *assemblage*, *community*, *guild*, and *swarm* have all been used to describe a large group of grasshoppers or locusts. Entomologist Jeffrey Lockwood has suggested that there is no *absolutely* correct term. But this doesn't mean that anything goes. "A term is correct," he argues, "if it accurately reflects the conceptual framework of the investigator and effectively communicates this perspective to others."[12]

He is no doubt right, but this inconsistency creates some dilemmas for those promoting entomophagy. Do we use the non-insect-eater scientific names or the insect-eater names? We would like to be able to quickly pick out the edible ones, but the scientific names might give greater opportunities for identifying new ones and for communicating across cultural boundaries.

What many people now think of as science emerged from non-insect-eaters in Europe in the seventeenth century. Biologists, who are part of this tradition, categorize life according to Domain, Kingdom, Phylum, Class, Order, Family, Genus, and Species.[13]

Arthropods (a phylum) are members of the kingdom Animalia, in the domain (a.k.a. empire) of the Eukaryotes. With their tough, chitinous external skeletons, segmented bodies, jointed feet, and open circulatory systems, arthropods are the most numerous and diverse animals on the planet. They were probably the first multicellular organisms to make landfall, preparing the way for plants. Insects — members of the class Insecta — are arthropods, but not all arthropods are insects. Besides the Hexapoda, the subphylum to which insects and a couple of smaller groups belong, arthropods include Crustacea (shrimps, crabs, lobsters), Chelicerata (spiders and scorpions), and Myriapoda (millipedes, centipedes, symphylans). People have eaten all of these, but this book is about insects, and I will refer to these other relatives only when they are important to the entomophagical narrative.

So a house cricket (*Acheta domesticus*) would be:
- Domain Eukarya (having a membrane-bound nucleus)
- Kingdom Animalia (animals)
- Phylum Arthropoda (arthropods)
- Subphylum Hexapoda (hexapods)
- Class Insecta (insects)
- Order Orthoptera (grasshoppers, crickets, katydids)
    (Suborder Ensifera [long-horned Orthoptera])
    (Infraorder Gryllidea [crickets])
- Family Gryllidae (true crickets)
    (Subfamily Gryllinae [field crickets])
- Genus *Acheta*
- Species *domesticus* (house cricket)

For the house cricket, this looks easy, even with the extra sub- and infra- groupings, but in general terms, the questions of naming and counting insects — activities which are closely related, since it's hard to count things if you can't identify them

— turn out to be a lot more complicated than they look, even for scientists. In some classifications, for some insects, there are superfamilies somewhere above families, and new genetic studies are changing the way bugs are classified. Scientific naming, then, is imperfect, but it's a place to start.

The terms *arthropod* and *insect* predate European non-insect-eater sciences. Pliny the Elder, who lived in the first century of this current era, was a Roman naturalist and army commander, who, in the ambitious manner of a military leader, sought to describe all living things. He made a great many statements about the natural world, some of which turned out to be right, and some quite wrong (such as the observation that caterpillars originated from dew on radish leaves). Among his legacies, he left us the word *insectum*, meaning "with a notched or divided body" or "cut into sections." Pliny's word was actually his translation from the Greek of *entomon*, which Aristotle, a few hundred years before, had used to classify our little segmented relatives, and from which we derive *entomology* and, more recently, *entomophagy*.

All of us are, in various ways, segmented, often both in mind and body, but in arthropods the sections (head, thorax, abdomen) are more obvious and more obviously specialized than in most other animal types. The non-extinct arthropods — the largest phylum in the animal kingdom — encompass arachnids (spiders, ticks, mites), myriapods (millipedes, centipedes, symphylans), and crustaceans (crabs, crayfish, barnacles, krill), as well as insects. All arthropods have external skeletons (exoskeletons) and jointed appendages, not all of which are legs. But one who is fascinated by legs could do worse than spend a few summer months standing on the corner watching all the bugs fly by.

The class Insecta encompasses about thirty orders. I say "about" because the number changed while I was doing research for this book. These orders include Coleoptera (beetles like dung beetles and Colorado potato beetles), Hemiptera (true bugs, like cicadas and bedbugs), Orthoptera (grasshoppers and crickets), Diptera (flies and mosquitoes), Hymenoptera (bees, ants, wasps), Siphonaptera (fleas), and Lepidoptera (butterflies and moths). One might think that more sophisticated science and more complete information — for instance, about genomes — would clarify and simplify our understanding of insects. Not so. A 2014 scientific article on the classification of some beetles, based in part on penis structure, was titled "A Preliminary Phylogenetic Analysis of the New World Helopini (Coleoptera, Tenebrionidae, Tenebrioninae) Indicates the Need for Profound Rearrangements of the Classification."[14] The authors spoke not only of the usual order, genus, and species, but threw in tribes, clades, and — just to make sure they'd fully covered the territory — polyphyly, polytomy, and paraphyly, which are classifications based on variations in genetically identified ancestors and descendants.

Insects account for more than 80 percent of all described living species. Among these, four orders predominate: Coleoptera, Diptera, Hymenoptera, and Lepidoptera. Although there are about a million named insect species, some researchers estimate that there could be another million — or several million — waiting to be discovered and named. There is also the counterpossibility that some of those already named are varieties of the same species. To put this into some sort of context, there are about 20,000 species of fish, 6,000 species of reptiles, 9,000 species of birds, 1,000 species of amphibians, and 15,000 species of mammals. So, most animals are arthropods, and most arthropods are insects. According to an oft-repeated quote, allegedly from paleontologist J. Kukalová-Peck, to a first approximation, every animal is an insect.

Which brings me back to the use of *beetles* in the title of this book. Why talk about eating beetles? Why not insects, or even bugs? True bugs, like bedbugs, cicadas, aphids, giant water bugs, assassin bugs, and scale insects, comprise "only" about 80,000 species of the millions of insect species out there. Despite this, some of the best books on insects and their taxonomic relatives written by entomologists have "bugs" in the title. These include May Berenbaum's *Bugs in the System*, Scott Richard Shaw's *Planet of the Bugs*, and Gilbert Waldbauer's *What Good Are Bugs?* The term *bug*, with Welsh and Germanic roots, was first used in medieval times to refer to devils, hobgoblins, ghosts, and other unseen and sometimes frightening annoyances, which may accurately reflect the kinds of arthropods encountered by medieval Europeans. Today, the word *bug* carries that etymological baggage and more, being used to refer to small insects, disease-causing bacteria,[15] hidden microphones, and computer glitches.

If we consider only insects that people eat, however, the picture changes. There may be 1,900 species eaten by someone, somewhere on the planet, but some families and orders — Coleoptera, Hemiptera, Hymenoptera, Isoptera, Lepidoptera, Orthoptera — are more popular than others. Lepidoptera (butterflies and moths, eaten as caterpillars), Hymenoptera (bees, wasps, and ants) and Orthoptera (grasshoppers, locusts, and crickets) each contribute somewhere between 10 and 20 percent of the global insect diet. Cicadas, leafhoppers, plant hoppers, scale insects, "true bugs," termites, dragonflies, and flies each contribute less than 10 percent. On a species basis, the largest insect contributors to human diets are the Coleoptera (beetles), which contribute about a third of the total number of species eaten. In many parts of the world, beetles are eaten as adults, but in North America and Europe, mealworms (baby darkling beetles) are the most popular.

There are more species of beetles (close to 360,000) than there are of all of the rest of us animals put together. If we are all, to a first approximation, insects, then I would add that, to a second approximation, we are all beetles. This also sounds like something a Douglas Adams character would say. The word *beetle*, which comes from Old English and once referred to a hammering tool, and then a "little biter," carries less baggage; that's one of the reasons I have used it in the title of this book. A German car, and then, misspelled, a musical group, are not likely to add confusion to the entomophagist's kitchen.

In the past few hundred years, non-insect-eater scientists have attempted to overwrite traditional insect names with Linnaean taxonomies. Yet it is not always clear how that insect-eater knowledge maps onto non-insect-eater science, or vice versa. Given that we have no clear idea how many species there are, and that many indigenous cultures are disappearing or being absorbed into some variant of the generic "global" insect-avoidant McCulture, however, it is easy to feel as if one is a prisoner wandering around in the luminous fog of George Lucas's pre–Star Wars movie *THX 1138*.

Once we get "down" past arthropods and insects, and into the crowded Victorian streets of species and genus, it becomes very clear that most scientists are from non-insect-eater cultures. About a million insects have been given two-part Linnaean names; I could find none that related them to food, and many are neither ecologically informative nor globally understood. Some, like

*Mantis religiosa* (the European praying mantis) reflect religious practices. Others are both whimsical and culturally specific: the second name of *Dicrotendipes thanatogratus* reflects the entomologist's love of the Grateful Dead, while others, such as a parasitic wasp named *Heerʒ Lukenatcha* and a fly named *Pieʒa kake* — you have to say these last two out loud — are clever insider jokes. Many insects are named after famous people, mostly from Europe or North America. Entomologists have named parasitic wasps after Ellen DeGeneres, Jon Stewart, Stephen Colbert, J.S. Bach, and Ludwig van Beethoven, for instance. In 2013, John Huber and John Noyes reported on a new genus and species of fairyfly in Costa Rica, which they named, in a flight of scientific fancy, *Tinkerbella nana*. Since fairyflies are parasitoid wasps who lay their eggs inside the eggs of other insects, thus killing them, this offers a somewhat sinister reframing of the Walt Disney–J.M. Barrie fairy Tinkerbell and the children's dog Nana.

One might be forgiven if one thought that all this naming and differentiation down into orders, families, and species was a case of Linnaean OCD, but the classification system is a reflection of both our language and the complex differentiations that occur in observed nature. Insects have evolved their own version of the "narcissism of small differences," a phrase that Freud used to describe the phenomenon of exaggerating small differences in order to emphasize one's own unique identity. For insects, these fierce differentiations are a feature that, along with their small size and high reproductive rate, have allowed them to fill the biosphere and make the world over in their own image.

For entomophagists, the ability to recognize fine differences between species is more than a quibble between specialists, postcolonially defined indigenous peoples, and postmodern entomophagists. As we consider the question of insect-eating, the distinctions enable us to more intelligently reconsider the contexts for environmental management, economic benefits, and human health outcomes. These distinctions are at the heart of which insects we see as invasive species, which ones we think will eat us the first chance they get, which will just annoy us, and which, even as they titillate our taste buds, might help us imagine a convivial future on earth.

Some of these differences will become clearer as we continue this journey into global insect-eating, but let me give a few examples here to illustrate the point. Members of the Tenebrionidae family, whose larvae are called mealworms by chefs, are also known as darkling beetles. The family includes beetles that live in chicken litter; chickens eat the beetles and larvae, but the insects can also eat the floor and walls of the chicken coops, and carry around bacteria and viruses that may infect the birds and those who eat them. Other members of the same family include confused flour beetles (yes, that's a real name), mealworm beetles, and others that eat and recycle detritus in forest floors. So when talking about darkling beetles and mealworms, some precision in naming is called for if we wish to avoid unnecessary battles between entomophagists, chicken farmers, and forest ecologists.

Control of the invasive citrus pest California red scale (*Aonidiella aurantii*) was delayed by six decades because entomologists had lumped together all the ectoparasitic wasps *Aphyrus* as if they were a single species. *Aphyrus* turned out to be a complex of species, some of which would attack the red scale. Similarly, fig production in California was held back by a decade until entomologists and farmers identified and imported one species among hundreds in a particular genus of wasps that would pollinate the trees. In Australia, when farmers brought cattle into the country in the late 1700s, an initial failure to differentiate dung beetles and to recognize their finicky, often species-specific eating habits, created disastrous bush fly and landscape management problems.

Until the 1970s, the mosquito *Anopheles gambiae* was thought to be the most important vector of malaria in sub-Saharan Africa. This "one" mosquito, however, was found to be a mix of seven different mosquitoes that looked essentially the same but varied in their ability to transmit disease and in their resistance to pesticides. When it comes to a disease that kills millions of people every year, the inability to precisely differentiate species can have tragic consequences.

Even using the word *entomophagy* is not without controversy, at least among academicians. Anthropologist Julie Lesnick has written that insectivory is appropriately applied to what non-human animals do when they eat bugs, and that *entomophagy*, a term which she does not like, applies to what people do when they eat insects. Since people are animals, and I am officially retired from academia, I will not take sides in the entomophagy versus insectivory skirmishes. Like Jeffrey Lockwood, I prefer to use a variety of words, as long as they communicate clearly. The guys at Entomo Farms refer to "a person who consumes insect protein in order to drastically reduce his/her carbon footprint and contribute towards the healing of the Earth" as a *geo-entomarian*.[16] For some reason, to my mind, that word conjures up the image of someone really old, like me.

Heather Looy and John Wood, in a thought-provoking 2015 article titled "Imagination, Hospitality, and Affection: The Unique Legacy of Food Insects?" suggest that "as we come to value insects for the vital roles they play in sustaining life, and as a source of food, our language will necessarily change. And if we want to stimulate this change, we need to discover new ways to speak, and therefore to think and enact insects as a dietary element. Speaking of 'insect-based foods' or 'edible insects' is certainly better than using the technical term 'entomophagy.' But we still lack a rich language of familiar and nuanced terms for meat from the taxonomic class Insecta."

I agree with Looy and Wood. We need a richness of language to match the diversity of this extended animal family of ours, one that draws on the millions of eco- and place- and cuisine-specific names indigenous peoples have conferred on our tiny cousins — some proto-Babel ur-tongue that includes and transcends entomophagy, gourmet insect-eating, and rustling up some (literal) grub. We'll need to include the languages that enable us to talk about the control of malaria and dengue fever, and to evoke the seductions of honey and the healing properties of termites. In this book, I shall be careful to use specific terms when those are warranted to make fine distinctions, but will use more general terms like *bugs* and *arthropods* when fine distinctions amount to overspecification, the linguistic counterpart of economists adding several numbers after a decimal point to give hard credibility to soft theories.

Our eyes, our imaginations, and our science tend to see — and to differentiate and name — the world in terms of what is immediately apprehended by our senses and what has been important for our immediate human survival and reproduction. Hence we see children (our genetic future), cows (food, fertilizer, labor) and tigers (a predatory threat) as individuals or small groups. But, with the exception of those that dig into our skin in sometimes embarrassing places, we tend to notice most insects

— ants, bees, black flies, mosquitoes, locusts, cockroaches — because of their sheer, overwhelming numbers.

Now that I have completely overthought, defensively over-explained, and taken all the fun out of the book title — and before I draw up a list of which insects we are inviting to dinner — let us spend just a few pages exploring the size of our extended six-legged family. Really, kidding aside: how many invitations are we sending out?

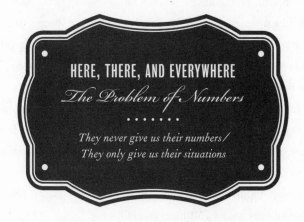

HERE, THERE, AND EVERYWHERE

*The Problem of Numbers*

. . . . . . .

*They never give us their numbers/*
*They only give us their situations*

Naming the contenders for our next chef's challenge is only one part of our entomophagical dilemma. If one argument for eating insects is that there is such an abundance of them, we might ask how many there actually are. And — this is particularly important if we wish to forage, manage, or farm them in some way — what are the characteristics that allow them to live in such abundance?

In 1691, British naturalist Sir John Ray, author of *The Wisdom of God Manifested in the Works of His Creation*, guessed at the proportion of insect species among all the animals in his homeland and projected this onto the world population. Using this most reasonable method, he came up with an estimate of ten thousand for the total number of insect species. In the eighteenth century, Swedish biologist and physician Carl Linnaeus created the binomial framework for naming living things. His initial descriptions included 4,023 species of animals, of which 2,102 were insects. By the mid-1800s, more than 400,000 insect species had been named, and by the early decades of the twentieth century, British entomologists David Sharp and Thomas de Grey had raised the total probable number of insect species to two million while American Charles Valentine Riley, considering the unknown unknowns living in the tropics, suggested that ten million insect species might be closer to the truth. Other estimates, based on expert opinions, or using proportions of all animals known to be insects and projecting them globally, ranged from a few million up to thirty million. Arguments among entomologists continue in the scientific literature. In a 2009 compendium on insect biodiversity, entomologist May Berenbaum simply subtitled her chapter "Millions and Millions."

Peer-reviewed scientific estimates since Berenbaum's publication have suggested that there are about 8.7 million species of all kinds on earth, of which about 90 percent have yet to be catalogued or described. The task given humanity in Genesis to name all things seems endless. In 2016, for instance, a group of researchers discovered and described twenty-four new species of assassin bugs. Considering the large uncertainties regarding all these estimates, Berenbaum's flexible "millions and millions" is as reasonable a place as any to start.

The next question is: given that there are so many species, how many *individual* animals might there be? Poet-naturalist Bill Holm imagined the population of boxelder bugs around his house by comparing them to adult male Norwegians and then imagining how much space they would take up.[17] Most scientists, however, use more pedestrian (although not necessarily more accurate) methods.

The UN has estimated that there are, at any time, about a billion cows and nineteen billion chickens on the planet, based on adding up numbers of documented animals country by country. In the same way, one could, theoretically, take each insect species, estimate the numbers for that species, and then add them up. That would assume, in the first place, that we knew how many species there were, and, secondly, that we could count the numbers for each of them and then add them up. I suspect, however, that our numbers in this case would tax even our most delirious statistical imaginations.

World-renowned biologist E.O. Wilson states (on the website of the Entomological Society of America) that "there are some 10 quintillion (10,000,000,000,000,000,000) individual insects alive." According to a 2016 estimate, there are about 7.4 billion people on the planet. If each "average" insect weighs three milligrams, then the total weight of bugs on the planet is seventy times the total weight of all the humans, assuming the "average" person weighs sixty kilograms. All of these are guesses, of course, since we don't actually know how many bugs there are on the planet. If you really need an answer more specific than that, I can recommend a good psychotherapist. Bottom line: there are a lot more of them than there are of us.

How did so many six-legged creatures come to be? Biologist J.B.S. Haldane once ascribed this abundance to an alleged creator's "inordinate fondness for beetles," but less fanciful reasons have also been proposed. In general, their numbers and diversity can be attributed to several interrelated characteristics, including being tiny, inhabiting many ecological niches, having multiple-use appendages, learning to fly, being sexually and reproductively clever and adaptable, (mostly) having lots of babies, discovering the joys of complete metamorphosis, and pretending to be leaves and flowers.

Let's look at some of these causes of abundance more closely. Apart from a few slow-reproducers, like tsetse flies, who give birth to a single larva every nine to ten days for a few months, most insects are what biologists call r-strategists. They produce a lot of young, thus increasing the chances that some will survive. Depending on temperature and feed, crickets, for instance, can lay ten eggs a day (a hundred in a lifetime); black soldier flies can lay hundreds of eggs in a five- to eight-day life, and ant and termite queens pop out tens of thousands per day (up to millions in a lifetime). The individuals in an ant nest may be numbered in hundreds of thousands, a termite nest in millions, and a locust swarm in billions. Insects are not only prolific baby-makers; some can also reproduce parthenogenetically — that is, females producing young without the intervention of males. Virgin birth can occur in some fish, birds, reptiles, and amphibians — and has even been rumored to occur in humans — but is most common in arthropods. For ants, bees, aphids, rotifers, and other insects, parthenogenesis may result in different kinds of offspring, requiring different food sources, which can enable them to adapt to more diverse climatic and ecological conditions.

Because insects are mostly very small and adaptable, they can occupy a great many ecological niches. Various insects and stages of insects feed on roots and parts of roots, fungi in the soil, buds, plant stems, flowers, fruits, the upper and lower sides of leaves, and on other insects that feed on leaves, roots, flowers, and stems. In the immortal words of Jonathan Swift, in his "On

Poetry: A Rhapsody" (which is actually long on parody and short on rhapsody):

> *So, naturalists observe, a flea*
> *Has smaller fleas that on him prey;*
> *And these have smaller still to bite 'em,*
> *And so proceed ad infinitum.*

Consider the Hymenoptera, for instance, which have played essential roles in the evolution of entomophagy and will surely be critical partners in any sustainable future. Xyelid sawflies evolved in the Triassic treetops a couple of hundred million years ago; they mostly ate pollen, buds, or leaves, but became the ancestors of millions of species of bees, ants, and wasps, including the fairy wasps sometimes referred to by their more endearing name of fairyflies (Mymaridae). There are 1,400 known species of these tiny (less than a millimeter in length) wasps. They are parasitoids, meaning that they eat other insects. Like many parasitic wasps, fairy wasps are important predators of plant-feeding insects — a natural and essential form of insect population control, which is important if we are serious about entomophagy. I shall come back to this later, but for now we are just looking at numbers. Fairyfly eggs are inserted by their moms into the eggs of other insects, which provide a fresh food source once they hatch. If you are very, very tiny, the world offers so many options for housing and food! There are also parasitoids of parasitoids, and parasitoids of those, all of which represent the converse of the old saw about a revolution devouring its children. In this case, the insect revolution's children devour their evolutionary parents. *Euceros* species, for instance, are hyperparasitoids of the larger Ichneumonidae (more than eighty thousand species) wasp family; the larvae are deposited on leaves, and attach themselves to passing sawflies, wait for a primary parasitoid to emerge from the sawfly, and then move into it and eat it.

Another contributor to the overwhelming insect population numbers is that, most annoyingly to larger predators, many of them learned how to fly early in evolutionary history, seeking out new places to breed and new sources of food. One of my most vivid memories from summer camp on the shores of Lake Winnipeg is the day when huge swarms of what we called fishflies, but were more accurately mayflies, lifted from the water in a cloud of biblical proportions (which was Christian summer-camp language for "big") and coated all the cabins, trees, shrubs, paths, and campers. What our counselors never told us was that the amazing, exuberantly flittering cloud of bugs was in fact an orgy, the long-legged males transferring spermatophores to their dancing virgin female partners, who then dropped back to the lake, where they dumped their fertilized eggs into the water before a fish or bird could eat them. Oh, had we only known! What lusty adolescent tales we could have imagined! Not believing in evolution, our counselors also never informed us that mayflies are among the most ancient of fliers, having first taken wing more than 300 million years ago. For 150 million years, insects were the only animals to fly.

More than 200 million years ago, insects invented (well, evolved) complete metamorphosis, in which an animal goes through several distinct and very different stages — think caterpillars, pupae, butterflies. For one thing, this means that the immature stages (larvae) don't look at all like the adults (moths), a problem which we have already identified for naming edible bugs. Ecologically, complete metamorphosis has other implications, perhaps the most important being that larvae don't compete with adults for the same food sources. In fact, sometimes adults don't eat anything at all. More than 75 percent of all extant insect species undergo complete metamorphosis. For entomophagists, complete metamorphosis provides a more diverse palate of possible options, but for those concerned with foods other than insects, this multiplicity of forms is more problematic.

Grape phylloxera (*Daktulosphaira vitifoliae*) is an aphid that nearly destroyed the French wine industry in the nineteenth century and continues to generate consternation and anxiety among vintners and oenophiles the world over. Phylloxera are members of the Hemiptera (true bugs). They are said to have somewhere up to eighteen forms, according to a classification based on sexual and dietary habits as well as whether or not they have wings. We could start anywhere in the reproductively clever tangle of their lives, but let's start with what one would think is the most straightforward distinction: males and females. Phylloxera are born without digestive systems from eggs laid on the underside of grape leaves, and then proceed to do what any energetic stomach-less males and females would do: they mate. Then they die, but not before the female lays one egg on the trunk of the vine. The egg eventually hatches into a nymph, who then clambers up onto a leaf, creates a gall by injecting her saliva, and then lays eggs in the gall without having had the fun of mating. In any case, the next generation of nymphs may travel to other leaves, or down the stems to the roots, where they inject toxins, suck sap, destroy grape vines, hibernate, and — having read the books on sustainability, thinking ahead and all that — reproduce parthenogenetically for up to seven generations. If the spring weather feels just right after a winter hibernation, they rise with the sap to the leaves. Or wander off to new plants. Or grow wings and fly to new plants.

Some insects, not wishing to go to the bother of actually changing form, are more imaginative in their adaptations. As a class, insects have developed multiple uses for what appear to be simple appendages. In *Planet of the Bugs*, Scott Richard Shaw reflects on insect legs. "Insects use their legs," he says "for walking, running, hopping, fighting, grasping food, tasting food, grooming their body, swimming, digging, spinning silk, court-ship, sound production, and even hearing." These multiple uses of jointed limbs are but one of many characteristics that have allowed these animals to diversify, to occupy so many niches and offer so many services to natural systems, including the provi-sion of food for humans.

As well as being small, reproducing like crazy, evolving multiple-use appendages, and undergoing complete metamor-phosis, many insects pretend to be something other than what they are, excelling in the arts of camouflage and mimicry much more than Navy SEALs or Bigfoot. We all know about stick insects, but have you seen the lichen-colored katydid, the leaf-litter mantid, the walking flower mantis, or the walking leaf mantis? Even if you had seen them, you likely wouldn't know it. They look like sticks, lichens, leaf litter, flowers, and leaves. If you are very, very patient, you may see them walk, or fly. That's usually a giveaway, which they know deep down in their DNA, so that if you are impatient and try to make them move, they usually don't oblige, having grown up with the story that their mother repeated until it was ingrained in their DNA. "You know your grandpa on Mom's side? He moved. And then he was eaten."

Finally, generating large populations is, in evolutionary terms, necessary but not sufficient. Unlike dinosaurs and trilobites, many insects survived major global catastrophes that wiped out other species (to the survivors belong the spoils). Going back a few hundred million years, into the Permian period, beetles have had the lowest family-level extinction rates of any species. This is attributed at least in part to their varied diets and the diverse ecological niches they inhabit. Insects are survivors.

For entomophagists, how these millions are organized is as important as sheer numbers. Some are more appropriate for foraging, or loose semi-management, while others lend themselves better to agriculture. Those insects that produce large numbers within closely organized societies — the so-called eusocial insects such as termites, bees, and ants — are much more difficult to farm intensively than those that have lots of young without being organized into colonies, such as crickets, silkworms, moth larvae, and flies. Bees, which are considered domesticated and are used as inputs for industrial agriculture, are in fact not so well suited for the economic box into which they have been stuffed, and they suffer for it.

The large number and diversity of insects, and their strategies for having achieved world domination, are impressive — and pertinent to which ones we might like to eat and how we might most efficiently harvest, or grow, them. But before we dive whole-bug into the mealworm bucket and make them part of our daily diet, we may wish to further determine whether eating bugs is good for us. Would persuading more people to eat insects result in a healthier population and a more resilient biosphere?

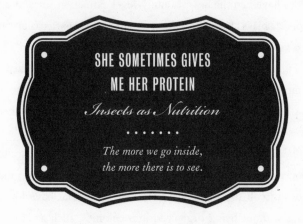

SHE SOMETIMES GIVES
ME HER PROTEIN

*Insects as Nutrition*

. . . . . . .

*The more we go inside,
the more there is to see.*

Recent popular and scientific literature repeats general claims that insects, as a source of human protein and energy, are at least as good as — and probably better than — other livestock, and in terms of their ecological and social impacts, much better. If one examines the bases of these claims, however, the evidence is more ambiguous.

In 2010, FAO published a collection of papers from a workshop that was held in Thailand in 2008, titled *Forest Insects as Food: Humans Bite Back*.[18] The author of one of the report's chapters repeated a claim, first published in 1960, that whole bee combs, brood (babies) and all, "come close to being the ultimate food and health supplement in terms of calories and a balance of carbohydrates, proteins, fats, minerals, vitamins." In another chapter of the same report, an author states that dried silkworm pupae (*Bombyx mori*) have calcium levels equivalent to milk or peas, and might be an alternative source of calcium for people who are milk-intolerant, or in countries such as China where dairy products are not often part of the diet. The author then claims that, with about 50 percent protein and 33 percent fat, it is "commonly said that three silkworm pupae are equivalent to one hen's egg." These statements about the nutritional value of eating beehives and silkworm larvae may be true, but if we are going to incorporate them into a global food strategy, we might wish to have evidence that is slightly more reliable than what is "commonly said."

Much of the enthusiasm for the nutritional value of insects has been based on anecdotes, or on single studies that have not been replicated. The scientific quest since the seventeenth century has been to attain a state of knowledge about the world in which everything we claim to know is falsifiable; that is, we should be able to design a research study that could disprove the claim. If the claim survives many efforts at disproval, we accept it, at least tentatively, as being true. The kinds of studies that are considered acceptable in the conventional scientific world must follow a set of strict protocols. Because of this, many scientists have learned to be skeptical of indigenous or anecdotal claims. For instance, until the claims of effectiveness for traditional Chinese medicine could be replicated under controlled conditions, in which results of treatment were tested against placebos or other treatments, they were often summarily dismissed as folk tales. Similarly, promoters of fermented wasp shochu (made from Asian giant hornets) in Japan tell stories of how drinking it will improve one's skin and cure fatigue. The stories may be true

— or true sometimes, and in some places — but until they are tested under controlled conditions, scientists tend to be skeptical. The effects we hear about in these stories may be "just" placebo effects — people getting better, or losing weight, or being more fit because they *thought* the medicine or the food was good for them, or because social circumstances quite apart from the medicine or food were what actually produced the effect. The so-called Mediterranean diet, for instance, may promote one's health not because of some magic chemicals in the wine or olive oil, but because the social context in which many Mediterranean people eat — with friends, at a leisurely pace — may be conducive to reducing stress and promoting health.

Having said that, we may irretrievably lose a lot of important information if we simply and arrogantly dismiss any and all anecdotal claims as "unscientific." It may not be that a specific chemical in the wine, or the insect, is of value, but we can still learn a lot by paying careful attention to information passed down through many generations of naturalists and healers. Without some replicated evidence that they are "bad" for us or for the planet, we are fools to simply dismiss traditional cures and stories.

The lack of replication is not peculiar to entomophagy; as formal reviews of the scientific literature in the twenty-first century have discovered, many scholarly studies — including many of those assessing the effectiveness of drugs we use in our medical system — have never been replicated. What this means is that the humility and uncertainty with which we approach entomophagy's claims should be applied to all claims of cures and panaceas, including those endorsed by scientists.

One of the ways to come to grips with this is to look at a variety of sources of information, to see if they agree. This is called triangulation, and it is one of the better ways of drawing on multiple, very different, sources of information. We may not be able to conduct a randomized clinical trial on claims of sustainability and health, but we can have greater confidence if the information we are using is verified by several sources and from a number of different perspectives. In doing this, we need to pay careful attention to some details. Are the reports of nutrient content on a dry-weight basis (which makes it easier to compare across species of animals — crickets to cows, for instance) or on an as-eaten basis (which allows us to better assess how nutritious the food is on the plate)? Was the research based on using different laboratory tests and under different conditions? Does the nutritional content of insects vary from season to season and ecosystem to ecosystem? While we can draw some general, and useful, conclusions about the value of bugs as food, we should certainly back off from making grand statements of unwarranted precision (decimal points in food values always make me suspicious).

Recognizing these problems, some authors have attempted to formally bring together information on the nutritional value of insects. In the absence of specific breeding and feeding programs, we would expect this information to remain stable. Certainly if we wish to make food-guide recommendations, we must assume this level of stability. Even for animals that have been intensively manipulated, such as chickens and cows, generalizations about protein, fat, and micronutrients are more or less stable over time, although these vary by breed, diet, and cut of meat. This leads us to expect that we might see general patterns, and variations within those, among insects according to species and feed sources as well.

A 1997 review by Sandra Bukkens at the National Institute of

Nutrition in Rome seems to have set the baseline and the tone for pretty much everything that followed. Bukkens concluded that, in general, as a group, insects "appear to be nutritious. They are rich in protein and fat and provide ample quantities of minerals and vitamins. The amino acid composition of the insect protein is in most of the cases better than that of grains or legumes and in several cases the food insects may be of importance in complementing the protein of commonly consumed grain staples among indigenous populations."[19] In the two decades between 1997 and 2017, the number of reports, claims, and original research studies has multiplied faster than a room full of bugs. Most would concur with Bukkens.

Julieta Ramos-Elorduy, a long-time researcher into, and advocate for, edible insects in Mexico, has published some of the more carefully designed studies. She and two other researchers reported in a 2012 paper that the edible Orthoptera (chapulines) of Mexico that they sampled had protein levels ranging from about 44 to 77 percent (note the big range). These protein levels were similar to soybeans (about 40 percent), eggs (about 46 percent), beef (about 54 percent), and chicken (about 43 percent), but lower than fish (about 81 percent). The chapulines contained more energy on a per weight basis than vegetables, cereals, and meats, with the exception of soybeans and pork. "The quantity and quality of the nutrients [of] these edible orthopterans," the authors asserted, "provides a significant contribution to the nutrition of the peasants who eat them."[20]

Much of this material sounds promising, although the claim that eating insects was good for "the peasants" makes me uncomfortable. Just the peasants? Prospective entomophagists, walking into the data on the nutritional value of eating insects, can feel as if they have stumbled into the Fire Swamp in *The Princess Bride*. How should they interpret all these numbers?

Fortunately, I had not quite sunk irretrievably into the Fire Swamp's lightning sand when, at the end of 2015 and the beginning of 2016, I was rescued by two rigorous scholarly reviews that brought together and evaluated all the information available (at least in English) on the nutrient composition of edible insects.

In a 2015 paper, Charlotte Payne and her colleagues from the UK and Japan asked the question, "Are Edible Insects More or Less 'Healthy' than Commonly Consumed Meats?"[21] To answer this, they conducted a systematic review of the scholarly literature and focused on six categories of edible insects:

1. Species gathered from the wild and sold commercially as human food: *Vespula* spp (wasps) in Southeast Asia; *Macrotermes* spp (termites) in sub-Saharan Africa.

2. Agricultural pests harvested for human consumption: *Encosternum* spp (stink bugs) in Southeast Asia and sub-Saharan Africa; *Oxya* spp (rice grasshoppers) in Southeast Asia.

3. Traditionally wild-gathered species, sold commercially as food, for which farming methods are being developed: *Rhynchophorus phoenicius* (palm weevils) in Asia, Africa, and Latin America; *Oecyphylla smaragdina* (weaver ants) in Southeast Asia.

4. Species successfully reared on a large scale and sold for both domestic consumption and export: *Acheta domesticus* (house crickets) worldwide; *Gonimbrasia bellina* (mopane caterpillars) in sub-Saharan Africa.

5. Species with a long history of domestication by humans

and also sold commercially as food: *Apis mellifera* (honey bees) and *Bombyx mori* (silkworms).

6. Species not traditionally consumed by humans that are currently farmed on a large scale, intended for use as food and feed: *Tenebrio molitor* (mealworms) and *Hermetia illucens* (black soldier flies).

Having listed the criteria for which insects they wanted to consider, the authors then set inclusion criteria, which required that the studies were based on whole, unfortified, uncooked insects, tested at a stage in which they were usually eaten — adults, pupae, larvae, or brood. They also wanted the results on an as-eaten, not dry-weight, basis. Since a lot of animal feeds are prepared on a dry-weight basis, this criterion excluded a lot of studies. In the end, they found no usable data for stink bugs, wasps, and rice grasshoppers.

Given the criteria, their review would have excluded a 2010 report from Japan that found that silkworm pupae were about 56 percent protein by dry weight (compared to the 18 percent on an as-eaten basis reported by Payne et al) and had an amino acid profile that would satisfy WHO requirements for proteins appropriate for human consumption.[22] The author of the 2010 report concluded that silkworm pupae "are good sources of high quality protein and lipid and have an α-glucosidase inhibitor, DNJ, which may retard the absorption of carbohydrates and reduce post-prandial hyperglycemia."

Payne's review found considerable variations within and between species. For instance, although the protein content of house crickets and honey bees was about 15 grams per 100 grams of edible tissue, and for mealworms 21 grams per 100 grams, in each case the variations around those averages were 8–9 grams per 100 grams. Similarly, weaver ants were about 11 grams of fat per 100 grams of insect, and the mealworms were about 15 grams of fat per 100 grams, but in both cases the variations were so large that I am unsure what "average" means. Since these were whole animals that were tested, the researchers had expected less variation than they found.

Protein and fat contents differ somewhat between pigs, chickens, and cows of course. Some of this variation is based on their genetics, what they are fed, and how they are managed, but much of it comes down to what is being tested: hamburger versus lean steak versus liver, or eggs versus thighs. Since variation among cuts of meat is not relevant to testing insects, which are eaten whole, Payne's review suggested that the variation for major components such as proteins and fats was mostly an arte-fact of nonrandom sampling and small numbers of insects sampled. It may have also been that these insects, not having been genetically selected for their nutritional values, actually do have large variations of protein and fat content.

Despite the variations in study type and results, Payne and her co-researchers did venture to make a few generalizations: palm weevil larvae and termites are high in saturated fats, while crickets and silkworms are relatively low. Most insects, espe-cially honey bees, termites, weaver ants, and palm weevils, were sufficiently high in iron that they could be "recommended as a food source to combat iron deficiency." Similarly, crickets and mealworms, being high in zinc, could be combined with iron to protect against iron deficiency. High copper levels in termites, palm weevils, and mealworms could be problematic, as could the high saturated fats in palm weevils and termites. Variation in micronutrients such as iron and calcium reported for insects may reflect genuine variation based on small sample sizes (think of comparing six haphazardly met people in a shopping mall), differences in geographical origin, and what the insects were fed.

A second literature review, by Verena Nowak and her colleagues at FAO in Rome, was focused more on gathering data for an international database and, as far as I could tell, cast a wider net, but summarized and reported results on only one insect — *Tenebrio* (mealworms).[23] Again, the study reported the nutrient content in grams per 100 grams of "edible portions."

Nowak's data on mealworms showed protein levels ranging from about 14 to 22 grams per 100 grams, which is within the range reported by Payne and her colleagues. Nowak's numbers on fat ranged from about 9 to 20 grams per 100 grams, of which 20–60 percent were polyunsaturated fats. Nowak and her colleagues further stated that according "to the thresholds for food labels [set by the World Health Organization and FAO] . . . *T. molitor* larvae are a source of calcium, zinc and high in magnesium; pupae are a source of magnesium; and adult mealworms are a source of iron, iodine, and magnesium, and high in zinc."

They also noted that these conclusions could be influenced by the practice of "gut loading," which involves "feeding nutrient-dense feeds in order that the nutrients [such as calcium] in the gastrointestinal tract complement the nutrients contained in the insect's body."

When we eat insects, as when we eat arthropods (including shrimp and shellfish), we are, with a few exceptions (such as lobsters), eating the whole animal — gut, shit, and all — so that gut loading can strongly influence the final products in terms of both nutrition and food safety. Shellfish are often put into clean seawater for a few days to flush out impurities, a process called depuration. Similarly, among societies where insects comprise an important part of the traditional diet, wild-caught terrestrial insects are often cleaned, de-gutted, and boiled before eating.[24]

Other original research papers describe mopane caterpillars as good sources of all the essential amino acids, linoleic acid, alpha-linolenic acid, and many essential trace minerals critical to normal growth, development, and health maintenance. Like many larval forms of insects, the mopane caterpillars are generally reported to have higher fat and protein content than the adults, and seem to compare well with chicken and beef. Payne and a different group of researchers also studied trace minerals in insects sold at market in South Africa.[25] They discovered that the mopane caterpillars had a surprisingly high level of salt, especially as sold in markets, where they were measured as high as 2,600 milligrams per 100 grams. The salt content of mopane caterpillars and manganese in termites led the authors to caution that "salt should be limited in commercial products; and that further research is required to determine whether common serving sizes of termites may put consumers in danger of manganese poisoning."

Although many studies have demonstrated the laboratory-determined nutrient content of insects, there remains the question of whether those nutrients are in a form digestible for humans. In a December 2015 interview on the BBC, Oxford biologist Dr. Sarah Beynon wondered whether Europeans, not having eaten insects for a very long time, might have lost the ability to digest them in ways that make their nutrients available.[26] The inability to digest chitin, for instance, has been mentioned by some skeptics as an impediment to the availability of proteins, fats, and micronutrients from insects. Insect exoskeletons are built up out of chitin, the second most common biopolymer (after cellulose) in the world. Is chitin, like cellulose or lignin in plants, just indigestible baggage — the suitcase the bug lives in, if you will — that comes along with eating insects?

Newer nutritional claims have not been as well investigated as the protein, fat, and micronutrient content of commonly eaten insects. If we wander into the shopping malls of diet and health claims and counterclaims, for instance, we come across reports that chitin is not only an important source of fiber, but also an essential part of a healthy, cancer-free life.

These claims about chitin could be dismissed as fantasy if there weren't a number of scholarly reviews reporting on the antioxidant, antihypertensive, anti-inflammatory, anticoagulant, antitumor, anticancer, antimicrobial, hypocholesterolemic, and antidiabetic effects of chitin and its derivative chitosan. Much of this research was done in the lab and in vitro, and focused on chitin that had somehow been processed, but if replicated in other contexts it would open up new possibilities for medicinal insect products that go far beyond food security.

One partial answer to Beynon's question about digestibility comes from a 2007 research report documenting the presence of chitinase, the enzyme that enables an animal to digest chitin, in the gastric juices of twenty of twenty-five Italian patients being examined at a medical clinic in Padua.[27] Although 5–6 percent of healthy Caucasians do not seem to have the ability to digest chitin, levels of chitinase are reported to be much higher in people from sub-Saharan Africa, particularly those who live in poor socioeconomic conditions. The presence of the enzyme not only speaks to the issue of digestibility today, but is also one piece of evidence from which some have inferred that early humans probably also ate insects.

Even if chitin is not fully digestible, the research literature suggests that it could be a very useful source material for high-tech insect processors like Ynsect. In a similar biotech vein, in 2016 a team of laboratory researchers reported on a study of the "milk" produced by *Diploptera punctata* (the Pacific beetle cockroach),[28] which — like the tsetse fly — gives birth to live young. Protein crystals in the "milk" of these beetles have three times more energy content than the same amount of buffalo milk, which is richer than cow's milk. Buffalo milk is used for ghee in India and mozzarella cheese in Italy. I suspect that we are not likely to see pasture-fed cockroaches lined up in a barn for milking, nor a worldwide craze for cockroach mozzarella, but this work does open up the prospect of new bioactive insect-based products.

Given all the scientific cautions about the variation and uncertainty of measurements, are insects as good for people to eat as the new entomophagists suggest? It depends. These data, and the wide variations reported, should give us pause when we're making general claims about the superiority of insects. Payne and her colleagues, recognizing these uncertainties, suggest that honey bees, termites, weaver ants, and palm weevil larvae could be added to diets to combat iron deficiency, especially if complemented with house crickets and mealworms, which are relatively high in zinc. Finally, while termites and palm weevil larvae are packed with energy and protein, they also have relatively high saturated fat contents and may thus be less than ideal as primary food sources in populations among whom cardiovascular disease is a problem.

When making health claims, the ingredients are of course important, but not all eating is about the list of ingredients. Indeed, most of the benefits of foods have to do with the social contexts in which they are eaten and the ecologically complex relationships connecting how they are grown, processed, and transported from the land where they are raised to our mouths. The notion of fortifying local foods with micronutrients and vitamins, or of using nutrient-dense garnishes, has a well-established role in almost all cultures, as does the practice of eating seasonally appropriate foods. The Amazonian Tucanoan people, for instance, have traditionally eaten insects in quantities inversely proportional to the availability of fish and game; used in this way, entomophagy has made a significant contribution to stabilizing their protein intake.

If done well, working with traditional cooks and local food producers, this combination of garnish and supplements may combine the best of the ingredients with food preferences and nutritional concerns, creating an insect-based version of the Mediterranean diet. Some innovative chefs in Kenya, Nigeria, and Mexico have already enriched maize flour with termites and baked bread fortified with African palm weevils, wheat buns enriched with termites, and maize-based flatbreads enriched with ground mealworms.[29]

For some of us, however, our concerns have less to do with our personal health and more to do with whether these foods are healthy for the planet our kids and grandkids will inherit. Sometimes, those outcomes are in conflict with each other.

## OB-LA-DI, OB-LA-DA
### *The Last Green Hope?*
. . . . . . .
*Happy ever after in the market place*

The Beatles' bouncy, singable tune "Ob-La-Di, Ob-La-Da" paints a scenario in which romance, happiness, domesticity, family life, and gender relationships are worked out in lives based on a local market economy and music bands.

In an April 2016 *Motherboard* article titled "How Eating Insects Empowers Women," writer Matt Broomfield asserts not only that raising and selling insects empowers women, but also that "ten kilogrammes of feed produces six kilogrammes of edible crickets, but just one kilogramme of beef" and that insect farming "creates just 1 percent of the greenhouse gases generated by farming an equivalent mass of beef or pork."[30]

This is the ideal, the aspiration, of the new entomophagists. Insects will no doubt make interesting and important contributions to the sustainability and diversity of human diets around the world, but how those contributions play out in real time will be more complicated, I fear, than the glossy promotional literature suggests. Having accepted that insects are at least as good, nutritionally, as the best of our other meat options, let's explore the ecological aspects of this utopian bug-eating paradise further.

First of all, it's important to note that most of the social and ecological arguments in favor of entomophagy are based on farming, not foraging, insects. The basic argument is that a global shift from farming and eating the usual suspects (cows, pigs, chickens) to eating insects (crickets and mealworms, mostly) will decrease the human ecological footprint and mitigate climate change impacts, even as it provides sustainable food security for seven or eight or nine billion people. These assertions sound very attractive. The question is: are they true?

One way to begin to disentangle the probable reality from the improbable claims is by analyzing pieces of the problem and hoping that one can fit the pieces back together. That's the conventional way of doing science. It doesn't always work, but it is a place to start.

Often, when people are advocating for one kind of meat over another within a farming system, they use a measure called a Feed Conversion Ratio, or FCR for short, which allows for comparison of how many kilograms of feed it takes to produce a kilogram of steak versus how many it takes to produce a kilogram of chicken or cricket. A higher number means you need more feed to produce the same amount of output — meat, milk, eggs, crickets, and so on. A 2015 study[31] compared the FCR for *Acheta domesticus* (domestic house crickets) fed poultry feed or food waste with the FCR for carp (the fish, not the diem), chicken, pork, and beef. If we look generally at how many kilograms of dry feed it takes to produce a kilogram of edible meat, then crickets, carp, and chickens are in the same ballpark — between 1.3 (crickets on poultry feed) and 2.3 (chickens on chicken feed); pork is at 5.9 and beef at 12.7. These data suggest that crickets are at least no worse than carp or chickens. But the published studies report FCRs that are as low as 5 for cows, 3 for pigs, and even lower for farmed fish. The problem is that FCRs vary according to the *quality* of what the animals are being fed. The higher the quality of the feed, the better (lower) the FCR. What if we focus on grain-fed farm animals — say, chickens versus crickets — and look at their efficiency at converting protein in the feed to protein in the meat? Here, chickens and crickets are about the same.[32] At what point, a skeptic might ask, does the cultivation and processing of crops for crickets cause them to lose their ecological advantage?

To begin to explore that, we need to look beyond FCR, which is not the only measure of how "green" insects are compared to other livestock.

Consider, for instance, the issue of greenhouse gas (GHG) emissions. Some insects produce greenhouse gases, and different ways of farming insects or foraging for them have different implications for GHG emissions. The question of whether raising insects makes a smaller contribution to the overall total than the production of larger livestock, such as cattle, remains. Food-related greenhouse gas production is not simply a function of adding up the number of cows and crickets and comparing their collective average fart and burp volumes.

The authors of a 2014 article on "food-demand management" argued that "it is imperative to find ways to achieve global food security without expanding crop or pastureland and without increasing greenhouse gas emissions."[33] Most of the major organizations that work on livestock agriculture, such as the International Livestock Research Institute, based in Nairobi, and FAO,[34] admit that even if livestock-rearing offers poor farmers a path out of poverty, and even if that livestock has become ecologically essential in many cultivated landscapes, the kinds of large-scale agriculture developed in European-diaspora countries is simply not sustainable. We don't have enough water. We don't have enough land. What few can agree on, however, is how to increase food supplies to meet the demands of increasing human populations while striving toward some reasonable global semblance of social and economic equity without at the same time destroying the planet. For the European diaspora to lecture their former colonies about what they should or should not eat seems at best uncharitable, and at worst, hypocritically cynical. Can insects save us from this quandary?

On December 11, 2015, during the COP21 climate change negotiations in Paris, the BBC posted an interview with chef Andrew Holcroft, of the Welsh restaurant the Grub Kitchen, and his partner, Oxford biologist Dr. Sarah Beynon.[35] The piece was titled "How Eating Insects Could Help Climate Change." The interviewer, BBC Persian's Sahar Zand, repeated claims made by others that the amount of greenhouse gas released by producing 200 grams of steak is the same as that released by producing 20 kilograms of edible insects, which is an order of magnitude more conservative than Matt Broomfield's assertion.

Greenhouses gases may be emitted by livestock themselves (as in cow burps or termite farts), or by the methods used to produce them (industrial, grain-fed versus free-range and pasture-fed). For animals being grown for food, the question is how much GHG is produced per unit of weight gain. Under at least some experimental conditions, insects appear to be the winners. These conclusions, however, need to be reexamined in the context of the production systems being promoted and the landscapes on which they depend.

Termites are foraged, not farmed, but they do produce GHG. Furthermore, termite populations are modified — usually increased — by practices of industrial agriculture and forestry. May Berenbaum, in an essay collected in her book *Buzzwords: A Scientist Muses on Sex, Bugs, and Rock 'n' Roll*, reviews the reported evidence on arthropod flatulence. She notes that insect methanogenesis predates human society by many millennia, and that insect flatulence is a well-documented phenomenon. The *General History of the Things of New Spain*, for instance, compiled by Fray Bernardino de Sahagún in the sixteenth century, tells of an insect whose breaking of wind was alarmingly stinky. Nevertheless Berenbaum focuses most of her review on the intense efforts to quantify methane production by termites beginning in the 1980s. A 1982 report in the journal *Science* by an international team of scientists estimated that the trillions and trillions of termites on earth were producing about 30 percent of the methane in the earth's atmosphere (trillions and trillions of teragrams).

We might ask how greenhouses gases produced by wild insects are relevant to those we are eating. Well, for one thing, the study reported that the amounts produced were increasing over time as termites took advantage of all that termite food produced by deforestation and agriculture (and, I might add, the production of printed journals).

Other studies suggest that clearing land for monocropping

agriculture and deforestation is decreasing termite habitat, resulting in a downward trend for termite-associated methane production. Initial reports making the case for this view were followed by a series of arguments and research reports using different research methodologies and/or drawing different inferences from the same data. The general consensus that developed over the last decade of the twentieth century seemed to be that the amount of methane produced by termites depends on where they live — termites from Amazonia have the highest rates of emission — and, as it does for all of us, on what they eat, with soil feeders taking the lead, and wood feeders (in Berenbaum's words) "bringing up the rear." By the late 1990s, calculations suggested that termites were down to an estimated 5 percent of total methane emissions, but that emissions from cockroaches, for whom people were increasingly creating comfortable urban habitats, may have been offsetting whatever gains were seen from changes in termite populations. I wish there were some simple quantitative calculation to resolve this, but we live in a complex world where feedbacks and unintended consequences are to be expected. If we take a systemic view of agricultural ecosystems, then *if* a food system based on mini-livestock results in less land-modification than a macro-livestock–based system (and I think there is a very good case to be made for this), we end up with less GHG emissions from the insect-based food system even if the farmed insects themselves produce more GHGs.

One of the ways to try to integrate and make sense of some of this information is to carry out what's called a full life cycle assessment (LCA), looking at resource use and gas production throughout the entire food production system. Unfortunately, LCA requires that the "system" have boundaries; one of the challenges we faced in the 1990s, when trying to assess the health of ecosystems, was where and how to set these boundaries.[36] This often depended on the goals of the LCA and what one was measuring, whether the health of human communities, or migratory birds, or insects, or water availability. Communities have political, social, and geographical boundaries. Water use can be assessed within watersheds, but what if we are drawing water from a faraway lake? In comparing resource use of different animal-based agricultural systems, one would ideally wish to look at everything: fertilizer inputs to grow feeds and other crops, shipping and processing of those feeds, how the feeds get to the farm, how they are managed at the farm, what happens to the products once they leave the farm, how they get to the consumer, how all this alters GHG production by wild termites, and so on. Such ideal comparisons are, for the most part, unattainable. In comparing resource use and GHG production for different species of livestock, such as crickets and chickens, it is often most convenient to make assessments within farms, from the point at which animals are born (the cradle) to the time they leave the farm (the gate).

In a cradle-to-gate LCA of two tenebrionid species, the mealworm (*Tenebrio molitor*) and the superworm (*Zophobas morio*), researchers at Wageningen University quantified the insects' global warming potential, fossil energy use, and land use. Mealworm production required more fossil energy than production of milk or chicken, but similar amounts to pork and beef. Overall, however, the authors concluded that, given that land availability was the chief constraint to sustainable livestock production, and that farming mealworms required less land than farming other livestock, mealworms "should be considered as a more sustainable alternative to milk, chicken, pork and beef." Although this factor was not addressed in that study, cattle, pigs, and chickens also take a lot more water to grow than insects.

Issues of land use, and the larger questions of habitat conservation, raise some tough questions about the relative merits of foraging versus farming. The 2015 Japanese NHK World documentary *Hungry for Bugs* shows an insect forager hacking open several trees, looking for larvae of the longhorn beetle. Watching this, I was uncomfortable. It seemed to me that a great deal of energy was being expended for minimal food rewards. In purely biological terms, were the fat and protein rewards worth the effort? More problematically, I worried about the destruction of habitat through foraging. When foraging is an occasional activity (such as eating cicadas when they emerge), or based on light human population demands, this is not a significant issue. But if insect-eating is commercialized and pulls a "sushi," mainstreaming itself globally with increased economic returns — and if this eating is based on foraging — will this result in serious environmental destruction? Is farming better? I'll come back to this.

The first, and probably most effective, step in enabling the positive impacts of entomophagy and minimizing the risks would be to stop exporting European and American styles of livestock agriculture to other parts of the world. By learning about, facilitating, supporting, and improving traditional entomophagical practices throughout the world, we can encourage genuine intercultural conversations about how people can eat sustainably.

Thai entomology professor Yupa Hanboonsong, one of the stars of global entomophagical research, asserts that it is not her intention to convert the world to entomophagy. "My mission is to preserve our farmland culture and domestic species," she says. "Unless we treat this subject sensibly, the bug-eating tradition will disappear from our society very soon either from the extinction of insects themselves or due to their declining popularity. And, sadly, our younger generation will only know hamburgers and fried chicken."[37]

Managing climate change would, in this narrative, be an unintended positive outcome of behaving in biologically sensible and culturally sensitive ways. I am reminded of the cartoon I send climate-change deniers who occasionally accost me. A scowling man is standing up in a hall where someone has just delivered a lecture on responses to climate change. "What if it's a big hoax," the man in the audience asks, "and we create a better world for nothing?"

Are the grand claims made on behalf of entomophagy warranted? If omnivorous humans shift away from eating the usual suspects to eating insects, will the human ecological footprint on the planet decrease? Possibly, but one of Sarah Beynon's arguments, which I have heard echoed elsewhere in the entomophagy community, gives me reason to hesitate. If entomophagy is to have an impact on climate change, Dr. Beynon says, insect-eating cannot be a novelty; it would need to become a standard part of the diet. The current agri-food system creates a global food culture run by economically and politically powerful corporations and the governments that serve them, rationalized on an unsubstantiated assertion that this is the best way to feed the world's populations. This globalized agri-food culture, framed in terms any livestock farmer understands — that an unidentified *we* are feeding an unnamed *them* — benefits the shareholders in those corporations. If insects are merely inserted into this system, we may simply be perpetuating a zip-line ride to hell in a handbasket.

The current wave of enthusiasm for entomophagy opens up some interesting alternatives. Can entomophagy be a strong stimulus for, and contributor to, mitigation of climate change impacts even as we provide sustainable and equitable food security for seven or eight or nine billion people? The issues I have raised here assume that people will be, or should be, *farming* insects. Traditional insect-eaters have most often relied on foraging in the wilderness to gather bugs. In a very crowded world, harvesting directly from the wild might help preserve wilderness areas — or devastate them. At the same time, farming insects raises for many the specter of industrialized livestock production with its economies of scale and negative social and environmental impacts. Are those our only options? What if we were to look at other possibilities? Where would we find different forms of inspiration, different examples?

Before we consider the variety of options available, or jump too hastily or drastically into changing human diets and agricultural systems with the hope of creating "a better planet," we would do well to step back and look at the importance of insects in creating and sustaining the world we currently inhabit.

# PART II. YESTERDAY AND TODAY: INSECTS AND THE ORIGINS OF THE MODERN WORLD

Insects were here millions of years before us and prepared the way for our arrival. They created us, and their DNA is part of who we are. Before we eat them, maybe we should have a conversation. But how? What are the languages of insects? In what languages can we converse? They are so different from us! Let us explore the magical mystery of the world that they created, and we now inhabit.

**I AM THE COCKROACH**

*How Insects Created the World*

. . . . . . .

*Bugs are us and we are bugs*
*and we are all together*

Sometime more than three billion years ago (give or take), life emerged on earth. During the next billion or so years, single-celled organisms lay back and just kind of bubbled around in the warm water, producing wastes such as oxygen and experimenting with carbon dioxide as a building material. About 2.3 billion years ago, atmospheric carbon dioxide levels dropped and the earth toppled into the first of several catastrophic ice ages. Over the next 500 million years, there were four global ice ages, including what has been called "snowball earth," around 610–690 million years ago. In between those ice ages, life on earth experimented with a variety of options, recreating itself in diverse and exciting waves.

Then, during a period of time more than 500 million years

ago, when the earth was spinning faster than it is now, the moon was closer, and young love was in the air, the adolescent con-. tinents of Gondwana and Laurentia, as well as Siberia and Euramerica, moved away from that (perhaps) mythical mother of all land masses, Rodinia (from the Russian *ródina*, for homeland). Ah, those were the days! At some point — if I might be allowed some poetic license to consider millions of years a point — Gondwana and Euramerica sidled up to each other; together, they formed one giant continent, Pangaea. But that happened later — like, we're talking hundreds of millions of years later — during what came to be called the Permian period, named for the Russian province of Perm Krai, near the Ural Mountains, where many rock strata from this time have been found.

Also during these pre-Permian days, single cells discovered the joys of exchanging bodily fluids, and what has been called the "Cambrian explosion" unfolded like slow-motion fireworks. The years from the Cambrian period (570 million years ago) to the end of the Permian period (250 million years ago) are sometimes lumped together into the Paleozoic Era. When evolutionary biologists describe events occurring over twenty-five million years as an explosion, we might be forgiven for pointing out that their understanding of an explosion differs from that of war refugees or gold miners. Stephen Jay Gould, writing about the almost hallucinatory variation in Cambrian fossils discovered in Canada's Burgess Shale, called this the era of "Wonderful Life."

Among the many living things that came into being over the next few hundred million years were the first attempts at achieving insect-like perfection. Secreting waste products that solidified, some organisms developed external protective gear. Ever since that time, those of us with internal skeletons have, with various sorts of shields and body armour, dressed up like jousting knights and riot police in an attempt to mimic the incredible success of exoskeletons.

Those early animals that survived to reproduce had multiple jointed legs. The happy campers in the warm Cambrian seas included *Anomylocaris*, a meter-long predatory arthropod with long, spiny feeding appendages that fed on (among other things) thousands of species of smaller trilobites with segmented, three-lobed hard skeletons and jointed legs. It is fairly easy — and lazy — to lump together all these arthropod ancestors. In fact, some scholars have labeled the time before 400 million years ago the "Age of Invertebrates," which is just another way of saying the "Age of Things That Are Not in Our Family." If we were to look for vertebrate human ancestors during those years, we would have to dig through the bottom sediments in search of a slithering, tiny Gollum, the soft-bodied worm-like animal we now call *Pikaia*.

Cambrian life writhed its slow orgiastic tangle through the Ordovician period (488–433 million years ago) and into the Silurian period (444–419 million years ago). During the scant thirty-odd million years of the Silurian, several types of arthropods were the first animals — and possibly the first living things — to colonize land. Some evolutionary biologists assert that arthropods were thriving on land for millions of years before plants arrived. The arthropods did not write the history of this living planet; as the first farmers, they were too busy making that history, preparing the soil for the tall, tropical Devonian trees. The myriapods — which included the predatory centipedes, the friendlier, mostly scavenging millipedes, and the most insect-like of the group, the symphylans — were among the first land-dwellers. With segmented bodies (like all arthropods), and two legs per segment, symphylans are very small (2–10 millimeters, which is less than half an inch) and, like millipedes, can live on various organic materials in soil. In those early days, these myriapods prepared the way for the much later entrance of pre-human vertebrates. The first recorded (terrestrial) insect fossil, *Rhyniognatha hirsti*, dates from about 400 million years ago. Our vertebrate ancestors, the sluggish lungfish, only crawled out of the swamp forty million years later, during the Devonian period.

In the broader entomophagy movement, we occasionally see insects lumped together with scorpions and their arachnid relatives like spiders, ticks, and mites. Arachnids are members of the subphylum Chelicerata: arthropods that cannot digest solid food and that have special appendages for grasping prey. They are related to insects and can be viewed as part of the ancestral arthropod family, but they are not as closely related as we once thought. In 2010, a team of American researchers led by Jerome Regier from the University of Maryland Biotechnology Institute concluded, based on a combination of genetic sampling and complex statistics, that terrestrial insects are more closely related to lobsters (for instance) than to millipedes or spiders. For those promoting entomophagy, these relationships have marketing possibilities. See that cockroach on your plate? Think *lobster*!

In this age of obsession with carbon use, carbon taxes, and projecting our narcissism onto the natural world, some hive off the late Devonian period and refer to it as the Carboniferous period. It could also be called the Age of Peak Coal, or — given that atmospheric oxygen went as high as 35 percent — the Age of Oxygen. The Carboniferous period was also when the keyhole amphibians came around; these were the first four-legged vertebrates to develop ears, so that (we guess) they could engage in whispering sweet nothings and all that other important foreplay. Or maybe entomologist Scott Shaw is right when, in *Planet of the Bugs*, he argues that "it's probably no coincidence that vertebrates developed ears at a time when arthropods were starting to make a lot of buzzing and fluttering noises in the forest." *Hear that, honey? That's lunch! Meals on Wings!*

Insects, as usual, were millennia ahead of Amazon.com with the idea that one might use drones to deliver their goods. In fact, in an insect-centric version of the evolutionary narrative, we should really call the Carboniferous period the Age of Roaches, whose thrillingly winged ancestors made up about 60 percent of known Carboniferous insects. Roaches have been given a bad rap by a few destructive city cousins, but many thousands of modern species live in tropical forests, in the leaf litter, in the canopy, in caves, in bromeliad water basins, in daylight and darkness, at dusk and dawn.

Among the flying air-dancers during the high-oxygen Late Carboniferous were the largest insects that ever lived, the griffenflies (Meganeuridae). *Meganeuropsis permiana*, one of the Permian descendants in this family, was a predator with a 71-centimeter (2.3-foot) wingspan, spiny front legs for grabbing, and very strong jaws — the better to chew you with. With no birds, bats, or pterosaurs to compete with, these air dragons must have had a Wild West shoot-'em-up of a time.

Which bring us, finally — and here I really do mean finally, as it was the last age for most living things on the planet — to the aforementioned Permian period (542–250 million years ago), which brought to a close the Paleozoic Era. The Permian period ended with the greatest mass extinction in the history of planet earth, but, my-oh-my, what a Titanic celebration that age was before it went down! Many of those species that eventually became food and singing entertainment for pre-human primates and postmodern people were breeding, reproducing, skittering, and scattering onto the world stage during the Permian. These arthropods included the Protorthoptera, ancestors to the crickets, grasshoppers, and katydids, our modern karaoke singers, plague-makers, and storytellers. They also included the first insects in which the transition from babyhood to adulthood (egg, larva, pupa, adult) involved dramatic changes in form and diet (that is, complete metamorphosis): the beetles, lacewings, caddisflies, and scorpionflies. Dome-headed, and later two-tusked, warm-blooded proto-mammals roamed the landscape; the smaller ones probably ate some insects. In fact, this was the first time (if tens of millions of years can be called a "time") that our mammalian ancestors (the proto-mammalian synapsids and "beast-faced" therapsids) and the ancestors of those whom we now wish to eat (the arthropods) shared the same landscape.

Although the number and variety of insect species still branched and flowered and tumbled across the landscape during Permian times, many of those who had dominated the aquatic milieu, like the trilobites, were already in decline. After having survived multiple ups and downs, adapting and thriving for more than 300 million years, the trilobites disappeared during that greatest extinction of life in the history of earth. Today, they have the dubious distinction of living on only in such names as *Aegrotocatellus jaggeri* (named after Mick Jagger) and *Arcticalymene jonesi* (named after Steve Jones of the Sex Pistols). Cockroaches, having abandoned the seas for terrestrial adventures, might have — had they the benefit of German culture — looked over their shoulders in *schadenfreude*.

The next couple of hundred million years, from about 250 million years ago to 65 million years ago, have been called the Mesozoic Era, comprising the Triassic, Jurassic (of dinosaur movie fame), and Cretaceous periods. During that era, the supercontinent of Pangaea broke into pieces, starting about 200 million years ago, as part of shifting tectonic plate cycles.[38] In the wake of the mass extinction and continental drifts, life on earth renewed, redirected, and reinvented itself.

Having been the first to make landfall and prepare the soil for plants and non-arthropod animals, arthropods went on from innovation to innovation. Xyelid sawflies, winged ancestors of a wild array of social insects — bees, ants, and wasps — as well as some, like the parasitoid Ichneumonidae, whom some might call anti-social, evolved in the Triassic treetops. If you are eating bee brood at a dinner party, and someone is quoting Solomon or Aesop and pontificating about the social lessons to be learned from insects, you might wish to remember that the eggs of saw-flies and wasps that are fertilized by males always grow up to be female, while unfertilized (virgin-birth) eggs are always males. In postmodern terms, if we were given to "learning from nature," we could perhaps frame this reproductive system in terms of female power and choice.

Over the next tens of millions of years, some of the arthro-pods that survived the Permian extinction went through a complex diversification dance with their bisexual, flowering plant inamoratas. We have been taught to envision this time as the age of the dinosaurs, but the world that the dinosaurs bumbled and stormed around in was replete with a wide diversity of insects that influenced, and in turn were influenced by, the larger beasts.

During the Cretaceous period (145–66 million years ago) many of the animals favored by entomophagical enthusiasts, as well as those bugs that have, through pollination and honey making, made possible the abundance of food in the twentieth century, coevolved with flowering plants. The dinosaurs roamed a world diversely abundant with Hymenoptera (wasps, bees, ants), Lepidoptera (butterflies), and many beetles (Coleoptera) and flies (Diptera). In 2006, entomologists reported finding a bee embedded in amber (in what is now Myanmar) that was more than 100 million years old.

Wasps, hornets, and yellow jackets are members of the Vespidae family, within the same order as honey bees. These bee relatives have, in recent years, been demonized by the general public, often because they attack people or, more often, because they attack honey bees, one of the few insects to be granted saint status in non-bug-eating cultures. But we should not be too hasty in attacking them. In the first instance, you are much more likely to be attacked by another person than by a wasp, so the general military case against wasps is here on shaky ground. But there are other reasons, based on an understanding of ecological relationships, to be more circumspect and selective in how we respond to wasps. Of the seventy-five thousand species of wasps living today, fewer than 1 percent attack bees; they have many other roles in natural systems.

The ancestors of our sweet, friendly, fuzzy, helpful, vegan bees were predatory wasps that emerged from the flower gardens of this Cretaceous Eden. Furthermore, if we are looking for early experiments in which evolutionary forces tried out new behaviors and technical skills to which we are now genetic heirs, the wasps deserve some consideration. A female nest-provisioning wasp, after stabbing a caterpillar with its antibacterial preservative (well before humans accidentally invented sulfa drugs and penicillin) and paralytic venom, will drag it into a nest that she has tunnelled or sculpted and lay an egg inside the still-living caterpillar's body. Some species will then scatter sand around the entrance to camouflage it, or drag a stone to cover the hole. Wasps were thus stone tool users and disease managers millions of years before primates came around. Did the ancestors of our primate forebears learn toolmaking from them and then pass it on? Is this knowledge embedded in some of the genes we share with insects? If we include genetic memory going back to life's origins, perhaps human toolmaking is but a clever variation on the survival strategies of life itself, and not a feature that distinguishes humans from other species.

But we need not deny the annoyances of hornets, wasps, and yellow jackets, or speculate on the evolutionary origins of human behavior, to defend the importance of the Vespidae. Many Vespidae serve essential functions in promoting sustainable food security. I have already touched on their importance in fig tree pollination. Historically, in many parts of the world, wasp and hornet larvae, as well as those of honey bees, have been foraged and eaten directly. Furthermore, by preying on other insects, they offer nontoxic alternatives to pest control, which we will need if entomophagy is to succeed.

The ancestral Orthoptera (our old friends the grasshoppers and crickets), an order which today includes some twenty-five thousand species, also survived the Great Catastrophe. A 2015 study, using a combination of molecular genetics and statistics, decided that these orthopterans could be divided into two so-called clades, members within each group coming from a common ancestor: Ensifera (crickets and katydids) and Caelifera (grasshoppers and locusts). The crickets who, in a tradition honored even today among humans, sing to find mates,[39] went their merry troubadourian way more than 200 million years ago (Triassic and later). Katydids, the flower children of the Cretaceous, were selected for their leaf-shaped wings, blending in with the leaves of flowering plants; as a confusing aside, the Mormon cricket is actually a katydid, and is not a member of the famous tabernacle choir.

When the dinosaurs — except for the warm-blooded, feathered ones we now call birds — went extinct 65 million years ago, many tiny, six-legged animal species survived. Grasshoppers, at eight thousand living species the most diverse of the orthopterans, made their dramatic, munching appearance post-dinosaurs, just as the grasslands were evolving. Also, a few small, insectivorous, shrewlike mammals (our ancestors!) squeaked through the meteorite-blasted bottleneck.

It was after this creative destruction of the biosphere that the interactions between species of animals that are recognizably human and those that are recognizably insect begin to take forms that are recognizably modern.

Primates made their first appearances about fifty or fifty-five million years ago. The lineage of primates that led to the grandfather who is writing this book probably diverged from the chimpanzee lineage five to seven million years ago. Primate species such as *australopithecus*, who might be able to photo-bomb our family picture and be mistaken for Uncle Bob with a hangover, showed up in eastern and southern Africa a few million years ago, give or take. Some early people-like species left Africa and wandered north and east, looking for the fabled spice islands of the Indonesian archipelago. Others stayed and evolved on home territory for a couple of million years, only migrating out in the tens of thousands of years ago.

What stimulated the transformation from armpit-scratching pre-human foragers to head-scratching, insect-eating *Homo sapiens*? In an ironic twist in the evolutionary story worthy of a Sophocles tragedy, it appears that just as we are beginning to promote entomophagy in a big way, we are discovering that insects created us.

**WILD HONEY PIE**
*How Insects Created People*

· · · · · · ·

*Picture yourself in a bug on a river*

Bugs have been on the menu — at least as a condiment — from the beginnings of primate history. As for many prehistoric events, the direct evidence is scant, but there are various clues from which we can infer an insectivorous diet.

Some of this history can be surmised based on the behavior and diet of today's primates, many of whom are insectivores. Smaller primates, having higher metabolic needs than great apes, get a greater proportion of their diet from bugs than their larger cousins. However, all great apes eat insects; in general, like people, they have been drawn to those more easily seen and caught, like large, sedentary beetle grubs or other larvae; or to social insects such as honey bees, wasps, weaver ants, and termites; or to those that arrive in handy mouthfuls, at least periodically, like locusts and caterpillars.

Termite fishing and ant-dipping take time, but with patience, persistence, and the right tools, chimpanzees seem to think the activities worthwhile, as do those folks out in slow boats or hip waders. Jane Goodall noted that David Greybeard and Goliath, two chimpanzees from the Kasakela chimpanzee community in Tanzania, created termite fishing rods by stripping leaves off twigs. Researchers suggest that these chimps eat insects for nutritional reasons (of course), and that their specific choice of which bugs to eat is influenced by seasonality. I'm guessing that the rewards of fishing with your pals include peace of mind as well as fats and proteins; that is, as in people, insectivory in chimps has cultural as well as biological benefits.

This is borne out by a report published in 2014 that looked at insectivory in eastern African chimpanzees.[40] Drawing on fecal analysis, behavioral observation, and measurements of insect abundance, the researchers found that Semliki chimpanzees in western Uganda, whose communities are known to be among the most insectivorous in Africa, selectively ate honey and bees from *Apis mellifera*, as well as weaver ants (*Oecophylla longinoda*). The researchers suggested that, while there were "ecological time constraints" biasing consumption toward prey that could be gobbled up quickly, the particular species selected by the chimps might have been culturally determined. It is no great shock, then, to discover other evidence that supports the assertion that early humans probably ate insects for complex eco-social reasons.

There is indeed reasonably good archaeological evidence that, for nearly a million years, early hominids (starting about five million years ago) in southern Africa used bone tools to dig in termite mounds. If we make a few imaginative but biologically plausible leaps across the millennia, we can point to ten thousand-year-old Cro-Magnon cave drawings of grasshoppers, in the Cave of the Trois-Frères in Ariège, France, as evidence that these ancestors were paying attention to insects.

Rock and bark drawings of honeypot ants by Aboriginal Australians, and evidence that they dried and stored insects in empty squashes, suggest ancient practices going back before documented histories. From there, it is but a few more millennia to other cultural artifacts that connect us to bug-eating in the last, say, seven thousand years. In Mexico, anthropologist Julieta Ramos-Elorduy reports that the practice of storing insects in ceramic pots goes back three thousand years.

In the words of entomologist Scott Richard Shaw, "the origins of human tool use, fine motor skills, manual dexterity, and ultimately the rise of human civilization are firmly rooted in our ancestral insectivorous diets." In fact, Shaw writes, "We may owe our very existence to social cockroaches. If termites weren't abundant, would primates ever have come back down out of the trees? I doubt it."

And once we were out of the trees? What then? Much of the information we have on prehistorical entomophagy is based on anecdotes by anthropologists and archeologists, and a few targeted research projects. We can make some conjectures, however, from the living traditions of eating termites, mopane caterpillars, locusts, and grasshoppers in Africa, beetles in the Americas, and witjuti grubs in Australia. Insect-eating practices in central Australia, Amazonia, and eastern Africa are probably ancient, forage-based, and seasonal. The seasonality reflects the close ties that all insects have with plant cycles, ambient temperature, and rainfall.

Research into the dietary habits of the Tucanoan Indians in the Amazon discovered that they consumed more than twenty species of insects (mostly ants and termites), and that the species and stages of insects they ate were similar to those consumed by other indigenous Amazonian populations. Jena Webb, a researcher into the links between ecology and human health, described to me how she was guided by a ten-year-old boy along a pipeline in the Peruvian Amazon. As they passed through a flooded forest, the guide, bare-chested and wearing only shorts, "jumped from the slippery pipeline to the spikey palm tree and dug his arm into the heart of the tree to extricate a very large beetle, which he had seen scuttling along the tree from the pipeline. The beetle did not have big mandibles but it was big, the size of a small egg. Black. Then he discovered that it was not the only beetle. So one after the other he proceeded to remove the legs with his teeth and put the live beetles in the only receptacle he had, which was a tiny pocket in his shorts. He collected about 6–7. He walked along the trunk to get back to the pipeline and we continued on our fishing outing. We caught a couple of fish, which he strung up on a stick and we headed back after an hour. Immediately upon arriving at home he gave the treasure to his mom who began sucking the juicy, yellow liquid inside the beetle out with her mouth. She shared one with her two-year-old daughter. The mother had an air of contentment and pride."

From the description of the beetle and where it was found, I suspect that this was *Titanus giganteus*, one of the largest beetles in the world. They are thought to depend on decaying root systems of tropical hardwoods, and hence they will likely disappear as good-quality tropical hardwood forests are cut down. The ability of the boy to quickly identify the beetles as food and his mother's reaction would suggest a long cultural history of opportunistic entomophagy. The anecdote also suggests that the research uncovering the complementary eating of fish and insects among the Tucanoans can probably be extended — not surprisingly — to many other people.

Eating termites is an ancient tradition in eastern Africa, and knowledge of insect behavior and harvesting, as well as the medicinal uses of termites, is widespread, indicating a long coevolutionary history. In the Ugandan countryside, where I was out looking at tsetse fly traps, a young boy explained to me the seasonality of termite-eating. Poking at the insect soldiers guarding a termite mound, he explained that he and his family gathered termites for food when the winged forms emerged from the mounds in large numbers. These winged forms are the sexual stage of termites, and their emergence is related to a mating flight. They are attracted to light then and are more easily caught than at other times.

A veterinary colleague of mine recounted to me a story told by her son, who injured his back while digging a fish pond in western Kenya. Local children brought him "a large bowl that was crawling with live, winged termites. The Mama said they would cure the strained muscles in his back. He sat wrapped in a lesso, eating termites while the children chatted and laughed with him — dipping their hands into his bowl to share this delicacy that is limited to a very short season each year. A few days later, Peter was recovered enough to rejoin the work on the permaculture team."

Contemporary non-insect-eating cultures live primarily in temperate zones, where extreme seasonal temperature fluctuations would historically have made foraging for insects an unpredictable food source. Perhaps more to the point, temperate regions also offered ready multipurpose alternatives, with larger mammals such as cows, sheep, and horses supplying food, labor, and later — as agriculture developed and barns were attached to houses — warmth. In fact, the development and spread of agriculture in temperate climates about ten thousand years ago would have changed human attitudes toward insects more generally; that is, insects would now have been seen to be a threat to more reliable, abundant, domesticated food sources rather than a source of food in themselves. There is some irony, then, in contemporary Europeans promoting insects as an answer to food insecurity; but, of course, the context is fundamentally different.

In temperate zones, the history of insect-eating tends to be either seasonal or rooted in certain cultures. Silkworm larvae, a by-product of silk production, have been eaten for millennia in temperate areas of Asia. Aboriginal people in the North American plains ate locusts seasonally and opportunistically for over four thousand years. By contrast, the history of eating worm-infested *Casu marzu* cheese in Sardinia is probably recent (that is, measured in mere centuries). The exception to these general patterns is the unique human relationship with honey bees.

Among all the insects that figure prominently in global entomophagical practices, bees and their relatives have the longest documented role in early human evolution. That relationship has much to teach us about how entomophagy might find its place, both economically and culturally, in this century.

Honey bees probably evolved alongside the first humans in the tropical landscapes of Africa and, later, Southeast Asia. At least one evolutionary narrative suggests that honey bees as we know them today are descendants of an ancient lineage of cavity-nesting bees that left Asia around 300,000 years ago and rapidly spread across Europe and Africa. Other evidence points to prehistoric hive-harvesting in Africa. This would probably have been seasonal, with bees nesting in trees broken by passing elephants, or in caves. Rock art going back forty thousand years in southern Africa, the central Sahara, Zimbabwe, Australia, India, and Spain depicts this harvesting. One of the most famous of these drawings is about eight thousand years old, in the Cuevas de la Araña in the municipality of Bicorp in Valencia, Spain. The androgynous figure in the drawing (called, by a man no doubt aspiring to take advantage of the Man of La Mancha's PR machine, the Man of Bicorp), is climbing a liana (rope ladder) and reaching into what looks like a hive of wild bees. Although these early drawings are often described by non-insect-eating researchers in terms of harvesting honey, the people in the pictures were probably gathering a mash-up of larvae, adults, brood cells, honey cells, propolis, pollen, and wax.

The close relationship between the greater honeyguide bird (*Indicator indicator*) and humans in southern Africa reflects a long evolutionary history. These birds feed on bee eggs, larvae, and pupae, as well as beeswax and waxworms (the caterpillar larvae of wax moths). Rather than quixotically — and perhaps suicidally — trying to raid active hives itself, a honeyguide peeps and pipes and flitters to get the attention of indigenous honey hunters. The bird then guides the humans to the colony, stopping periodically, chattering and spreading its white-spotted tail, to make sure the slow, earth-bound hunters don't get lost or distracted. The hunters smoke out the bees, open the hive with a *panga* (a broad-bladed knife, like a machete), and raid the honey and brood. Once the hunters are gone, the birds swoop in to eat the leftovers.

If this bird–human–bee coevolution is important for understanding the creation of modern humans, then paying more careful attention to the context and history of human–bee relations can offer insight into some of the characteristics that are important for sustainable human–insect relationships. An edible bug is not just a crunchy bit of protein. Bees produce, for their own use, honey, wax, and propolis. They also collect pollen and produce high-protein edible babies. As the archeological evidence makes clear, people have historically stolen many of these "products" (if baby bees can be called products), for multiple uses. The pre-Classical Greek poet Homer saw bees as wild warriors; in the *Iliad*, he describes the Achaeans as "like buzzing swarms of bees that come out in relays from a hollow rock." People have used beehives as weapons of war, but honey also has a long history of use for wound dressing (the effectiveness of which was recently confirmed with experiments and clinical trials). The Greeks, from Homer to Hippocrates, extolled the various health and aphrodisiac properties of honey. It was, in retrospect, one of several insights into the natural world that they got just about right — although I'm thinking that the aphrodisiac part requires further research.

Over several centuries during the Homeric age and later, humans learned to keep bees in human-devised hives rather than just raiding wild hives. Egyptian temple wall paintings dating back to at least 1450 BCE show stacked, horizontal hives made of clay or mud, and the use of smokers to calm bees, suggesting that bees were, by then, domesticated. Until 1851, however, when the Reverend Lorenzo Lorraine Langstroth invented box hives with moveable frames, there was no easy or nondestructive way for beekeepers to harvest honey from the hives. That is to say, until the late nineteenth century, harvesting the honey would often have involved taking the bees, brood, wax, and all. Thus when people speak of the influence of honey bees on very early human evolution, the image that should come to mind is not the sipping of the sweet nectar of the gods, but the crushing and eating of stinging insects and the theft of their food stores.

Although the nectar of the gods is of course important. Honey, which is the bees' energy source, is produced from sweet but watery nectar; bees make millions of flower visits, suck up the nectar, process it from sucrose to fructose and glucose, spit it out, and fan it to reduce the water content. Finally, when the sugars have been processed and the water content has been reduced, the nectar becomes what people call honey, and the bees seal up the cell with a wax cap. The wax, every kilogram of which requires five to ten kilograms of energy from honey to produce, is used to construct their living quarters, including cells for eggs and larvae, honey, and pollen.

If the nectar is insufficiently dehydrated, then the natural yeasts begin fermenting the sugars, creating mead, an alcoholic drink going back at least to old Egypt. Watery honey ferments naturally, so the drink was probably a serendipitous discovery early in human evolution — back in our hardscrabble cradle-of-life days — to ease the long trek out of Africa to the Far East, or into the frigid European lands of the Neanderthals. In describing drunkenness, the classical Greeks spoke of being "honey-intoxicated." Incidentally, honey mead was the first alcoholic drink I ever tried, when my older sister bought me a bottle for my twenty-first birthday.

In the twenty-first century, some mead brewers have given this heavenly drink a Paleo-dietary twist. To be true to the ancient traditions, they declare, one should create what they call "whole-hive mead." This involves throwing the whole kit and caboodle — bees, brood, wax, pollen, propolis, venom, royal jelly, and honey — into a pot full of boiling water, mashing it, and letting it ferment. As apiarist William Bostwick declared on the Food Republic website: "What Odin drank, and Beowulf, and Vishnu (*madhava*, or honey-born) and Zeus (*melissaios*, one of the bees) . . . probably had chunks of comb and a few stray bees in it. Which is why I had to kill mine. Historical accuracy is a cruel mistress."[41]

Besides its uses as an ingredient in traditional mead, bee venom — based on "research-legitimated" folk practices — is used for cancer therapy and to treat arthritis; pollen, based on high hopes and good stories, is hailed as a superfood; and propolis, which the bees use to firmly stick stuff together, to plug up holes and cracks in hives, and to gum up spaces less than about a centimeter between frames and boxes (much to the frustration of beekeepers), is promoted for its alleged medicinal effects.

People have used beeswax to wrap cheeses, to make lip balm, and for shoe polish. Its use in casting bronze statues goes back six thousand years, and its use in producing batik in Southeast Asia goes back a millennium or two. The Catholic Church reportedly uses 1,500 tons of beeswax every year for candles.[42] Beeswax has enabled us, more recently, to track historical and prehistorical human–bee interactions. In 2015, Mélanie Roffet-Salque and a multinational flock of sixty-four other researchers reported in the journal *Nature* on the prehistorical use of bees by humans. By studying traces of *Apis mellifera* beeswax lipids used to seal ancient pottery, they demonstrated that, in Neolithic Europe, the Near East, and North Africa, bee products have been used continuously from the seventh millennium BCE (nine thousand years ago). In their study, the researchers found no evidence of the use of bees north of the 57th parallel (around what is today northern Denmark) during Neolithic times, which the authors attributed to climatic and ecological constraints. Nevertheless, the inclusion of "whole-hive" mead in Viking lore would indicate that northern Europeans had some familiarity with *Apis* species.

This multiple-use and value-added strategy that we see reflected in the history of human–honey bee relations has been picked up and extended by some of the modern insect production companies. Ynsect, for instance, uses insects (in their case beetles) "to bioconvert organic substrates, such as cereal byproducts, and transform those insects into sustainable nutrient resource for agro-industries and bioactive compounds for green chemistry." Ynsect's CEO, Antoine Hubert, may see the company's strategies to develop multiple-use insects and value-added products as new. In evolutionary terms, however, they are but the latest manifestation of a long agricultural tradition that started with, and builds on, human relationships to bees.

From shambling primates, to *Homo before-us*, to Egypt, China, Greece, and Rome, insects have made us who we are and continue to teach us lessons on how to survive. Anthropologist Alyssa Crittenden, in her 2011 article on the importance of the consumption of bees and bee products in human evolution, declared that "the ability to find and exploit beehives with stone tools may have been an innovation that allowed early hominins to nutritionally outcompete other species and may have been a crucial energy source to help fuel the enlarging hominin brain."[43]

While human interaction with social insects such as bees offers insight into the promises and dangers of managing insects for our nutritional benefit, the influence of insects and insect-eating on human evolution — and how we might view them on the plate — is even more profound. In 2000, a vinegar and pomace fly, *Drosophila melanogaster* (commonly known as the fruit fly), was the first multicellular animal to have its genome sequenced. This was a year before scientists first sequenced the human genome. The choice of this fly was not an accident. Since 1909, when Thomas Hunt Morgan proposed that *Drosophila* be used for genetic studies, this diminutive insect — unfussy, happily reproducing every ten days — has been a mainstay of genetic studies the world over. It would be no exaggeration to say that all the astounding feats of genetic engineering by self-congratulatory scientists in the twenty-first century — from tomatoes that never rot, to therapies for congenital disorders, to malaria-resistant mosquitoes, to pesticide-resistant crops — would not have been possible without the hard work, relatively simple genome, and lack of rebellious instincts in fruit flies.

In a curious way, all the hard work that *Drosophila* have done for us has resulted in a better understanding of just how much we owe to them. About 47 percent of the *Drosophila* gene also shows up in ours. This is similar to the overlap we share with honey bees (44 percent). We emerged from a common ancestor; they are part of who we are.

What does this shared ancestry mean, in "real life"? Neuroscientists Nicholas Strausfeld and Frank Hirth recently compared the neurological decision-making centers of insects and mammals. They concluded that the brain circuitries that mediated behavior and choice in arthropods and vertebrates were so deeply homologous that they must have derived from a complex common ancestor. The insect–human similarities included everything from the ability to stand and walk, to attention deficit and affective disorders, to impaired memory formation. We are, at some very deep level, transcendent arthropods.

Crittenden's review of evolutionary history, as well as reports from the front lines of genomic studies, echo a more ancient narrative, told and retold for millennia in the African nursery where we humans first discovered our wiggling toes and identified them as belonging to us. The San people of the Kalahari tell the following tale of human origins: A bee was carrying a mantis — her enemy — across a river. Finally, exhausted, she left the other insect on a floating flower, but before she died, she planted a seed in the mantid. From this seed grew the first human. Infant humanity was nurtured in a woven womb of insects, flowering plants, and water.

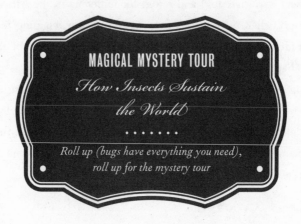

MAGICAL MYSTERY TOUR

*How Insects Sustain the World*

• • • • • • •

*Roll up (bugs have everything you need),
roll up for the mystery tour*

Given our blundering and destructive historical record in managing human–environmental relationships, particularly with regard to provisioning ourselves with food, we would do well to be cautious and scientifically astute before mainstreaming insects as food and feed. The ways in which our foraging or farming activities disrupt the life-sustaining activities of insects in the nonhuman world may well determine whether entomophagy can fulfill its potential for good, or whether it turns out to be disastrous. Even from a purely technical standpoint, an understanding of these complex ecological relationships is important for determining appropriate feed composition for farmed insects, improving their nutritional quality, and informing the debates between foragers, farmers, and those working for biodiversity and ecological sustainability. Eating insects is never *just* about insects and their value as food. It is about the web of life-sustaining relationships among insects, other animals, and plants.

Australian ecologist Tim Flannery, in his 2010 book *Here on Earth: A Natural History of the Planet*, concludes that if "competition is evolution's motive force, then the cooperative world is its legacy. And legacies are important, for they can endure long after the force that created them ceases to be."[44] We can approach this complex and cooperative legacy from a variety of angles, but it seems to me that all these approaches can be jostled into two broad categories. In the first is the question of *what* arthropods are doing — and what they were doing for the hundreds of millions of years before we humans got here for them to pester. In the second category is the question of *how* they are doing it.

In exploring these relationships, the *how* of insects — the ways in which they "see" the world and communicate — is as important as their more obvious and visible characteristics. It took centuries for people to begin to understand the sensory and hormonal world of domestic livestock and pets and its implications for management and welfare. As we begin to discern the lyrics of their pheromones and songs, the sensory and magnetic grammar of their languages, we can perhaps begin to discern more deft ways to manage our interactions with those insects that we consider pests even as we encourage insects and plants whose services and tastes we value. If this attentive understanding is the legacy of postmodern entomophagy, then it will have been worth far more than just discovering another food source.

Let's begin by looking at *what* they are doing out there. Despite our concerns with transmission of diseases like malaria, typhus, chikungunya, and dengue fever, insects are mostly really *good*. They recycle minerals and nutrients. They help plants (pollinating, dispersing seeds, supplying food, and providing defence) and animals (giving sustenance and protection). They limit population growth of plants, other insects, and vertebrate populations so that none of the rest of us, with the exception of *Homo moderna stultus*,[45] rampantly overpopulate the earth with reckless abandon; and they do the essential work of taking dead animals, dung, and plants and making them available for reuse by the living.

The work of insects in sustaining the living world is embedded in their dynamic relationships with micronutrients in the soil, with microbiota such as fungi and bacteria, with other insects, and with plants. Let us consider each of these relationships in turn.

The degree to which edible insects are sources of key micronutrients for humans is dependent in part on their roles in nutrient cycling in natural soil–plant systems.

Selenium is an essential trace element for all animals, from the very beginning of life billions of years ago. In living organisms, including people, selenium-based proteins are integral to protecting cells against oxidative damage. Some research suggests that it has a role in preventing certain forms of cancer. Selenium is one of those trace elements, like zinc, copper, and manganese, that in small doses are absolutely essential for life but in higher doses are toxic. It's an ingredient in shampoos to discourage dandruff and has other industrial uses. However, most of the selenium intake in human diets depends on the foods we eat and where they were grown. Its concentration and availability in plants depends on how it circulates in the soil and water.

And in many parts of the world, this circulation depends, in turn, on insects. Like other animals, insects need selenium, and by eating plants and each other, they accumulate it in their bodies; furthermore, they appear to have a higher tolerance for it than other animals. There are, as we have already established, very large numbers of insects, and, as they move about in the soil, air, and water and are eaten by other animals, they carry selenium from the plants, which draw it out of the soil, into terrestrial and aquatic food webs. They've been doing this for millions of years, thus helping to provide protection against oxidative damage to all living things.

Earlier, when I mentioned the major catastrophes at the end of the Permian and Cretaceous periods, I skipped over several lesser-known extinction "events" at the end of the Ordovician, Devonian, and Triassic periods. The events that ended the Permian (the biggest extinction of all time) and Cretaceous (when we lost the dinosaurs) periods lend themselves to our taste for heart-wrenching disaster, plus we have some reasonable ideas as to why they occurred (volcanoes, asteroids). Some of the other, smaller disasters have received less attention. In a 2015 article in the journal *Gondwana Research*, a team of researchers from Australia, Europe, and the United States reported that the three unexplained major extinction events on earth coincided with precipitous drops (in evolutionary terms) in selenium levels. The selenium levels they measured for these periods, which they took as an indicator of trace elements in general, were well below those thought to be critical for animals. Conversely, high levels of trace elements coincided with periods of high productivity, such as the Cambrian explosion. We don't know what roles arthropods had in these extinctions — although I am skeptical of the rumor that they hoarded the trace elements so that they could watch us die and then take over the world. What this means for entomophagists is that when we consider the food value of insects, we will also need to consider where the insects grew up and what they were eating.

The role of insects in the cycling and recycling of selenium gives us but a small hint as to their essential and multifarious role in sustaining life.

Consider termites, for instance. In semi-arid ecosystems, termites are considered a "keystone species" (that is, very important) for maintaining the soil. When my son and I drove a tiny Toyota Yaris across the partially flooded desert outback from Darwin to Adelaide, we stopped periodically to wonder about those giant, red earth termite skyscrapers rising amid the scrubby trees, built no doubt as an affront to primitive European invaders. In parts of Australia, as in the savanna and deserts of Kenya, Senegal, and Mexico, those termites — as they have for millennia — gather 500 to 1,000 kilograms of soil per hectare per year to build their community housing. This soil is then redistributed through erosion to the surrounding land. In some tropical forests, termites are the top detritivores, and may account for 90 percent of the insect biomass; in some places, data suggest that in addition to gobbling up half the leaf and grass litter, they eat and recycle 50 percent of the dead wood. I can think of a few offices where they might be helpful.


Termites are also an example of how insects evolved intimate relationships with the microbial world. Many people think that because termites eat wood, they can also, somehow, digest it. This is only partly true, and only for some types of termites. Termite guts are home to communities of bacteria, archaea, and, in some cases (in the case of what are known as "lower termites"), flagellated protozoans. These microorganisms depend on each other, in an arrangement called obligate symbiosis, to break down and digest lignocellulose, an important component of woody plant cell walls. Lignocellulose decomposes very slowly and is not very digestible by most animals.[46] In addition to termite microbiomes with protozoa and protozoa-associated bacteria being able to break down lignocellulose to access carbon, the termites need to get access to nitrogen in order to build amino acids and proteins. Some of this is accomplished through internal recycling, and some through a social behavior called proctodeal trophallaxis. This behavior, which involves seeking out and sipping droplets of hindgut fluids from nestmates (try not to think about it), can be traced back to the common, ancient ancestors of termites and wood-eating cockroaches.

About sixty million years ago, some termites (called higher termites by those who are on more familiar speaking terms with them than I am) lost their resident protozoa and were forced to get creative in how they managed their food supplies. Some of their descendants took up farming: *Termitomyces*, or the "termite mushroom," is so named because it grows only from termite nests, which can break down lignocellulose. According to a 2001 research report, so-called "old workers" forage for plant litter outside the termite nest. "Young workers" inside the nest chew up and swallow the plant material, but don't digest it before excreting it. They then take the fecal pellets and press them into a sponge-like "comb," which serves as a garden bed for the fungus to grow. As the fungus matures, its lignin content decreases, making the material more digestible. Young termite workers eat the fresh nodules while older workers eat the older mushrooms (which are more digestible). Sort of like young people eating indigestible granola for breakfast, and old folks getting fed porridge. Thus, termites are farming fungi, and in the process are making proteins and fats that are excellent foods for people and other animals.

Not all relationships between insects, bacteria, and fungi are so congenial. One might think about *Ophiocordyceps unilateralis*, the fungus that provides the premise for M.R. Carey's remarkable 2014 novel *The Girl with All the Gifts*. That zombie fungus, as some call it, invades the nervous system of ants in the *Campomotini* tribe, causing them to climb down from their canopy homes, settle on the underside of leaves, and then explode in a puff of spores. Or *Dicroceolium dendriticum*, the sheep liver fluke parasite, whose life cycle takes it from sheep dung to snails, and from snail slime to ants, where it takes over the brain. The ants climb up the grass, waiting for sheep to eat them to complete the cycle.

Moving up a scalar notch from insect–mineral and insect–microbe relationships, we encounter insect intrigues and betrayals worthy of a Shakespearean treatment.

Sometime during the Jurassic period, some wood wasps decided to give up their vegetarian diet and eat beetle larvae. Over the years, these carnivorous insects evolved and diversified from a few species to hundreds, to thousands, to hundreds of thousands. Rather than heading out into the woods and shooting down whatever they could get, these wasps, called parasitoids, developed a novel, more finessed, approach. Some of them, like the tiny fairyflies I introduced when we were puzzling over the millions and millions, inject their eggs into the eggs of other insects. Others, like the Tachinid fly family members (which are Diptera, or true flies) put their eggs just behind the head of a caterpillar, so when the eggs hatch, the larvae can eat their way into the living dinner. In some cases, a parasitoid wasp finds a young beetle larva inside a plant and injects a paralyzing, but nonlethal, venom into it. She then lays her egg next to the paralyzed larva. When the wasp baby hatches, it eats its way into the freshly preserved flesh. In a variation of this, some parasitoids insert their eggs directly into the insect that they want to parasitize; in this case, disabling the host's immune system is a challenge. Somewhere along the evolutionary way, a virus "decided" to hitch a ride with the egg, disabling the host's immune system to the benefit of both virus and parasitoid.

Some wasps have developed a cluster of cells called a trophamnion (a peculiar insect version of a placenta), which protects the wasp egg from the host's immune system and provides nutrition to the developing baby by absorbing nutrients from the host's blood. When the egg hatches — still inside the living insect host — it becomes what is essentially the tiniest aquatic insect, molting, floating around, sucking in food (but not defecating), and finally eating its way out into the outside world. By that point, of course, the host has been killed by the parasite. If the host is paralyzed, both the host and the parasite are stuck in one place. This works for the parasite as long as the host doesn't get eaten by some larger animal, such as a bird, or die for other reasons. In that case, both the host and its parasite are out of luck.

In the late Jurassic, another variation, called koinobiosis, emerged among the parasitoids; in this case the host, with the parasite inside, lives on and even goes through several molts before dying. In some cases, parasites of the parasites have emerged. In at least one case, a wasp that parasitizes the larvae of cecropia moths is itself parasitized by a smaller type of wasp, which may also be parasitized by an even smaller one — and, in the spirit of Jonathan Swift, that one is occasionally parasitized as well. These parasitoids are, in ecological terms, ingenious and diverse pathways for keeping various plant-eating and insect-eating populations in check, and they have been used for nonchemical pest control. They are doing us a favor then, so that we aren't overwhelmed by the bigger wasps and yellow jackets. They have also created wrenching problems for some scientists trying to imagine what sort of a god would have created such suffering — if one can call it that. Darwin's reluctant atheism (if indeed it was atheism) can be blamed on these creatures.

An important lesson for insect-eaters to take away from inter-insect relationships is that, in the wild, they keep each other's populations under control. Disturbing these relationships through the direct or unintended collateral damage of foraging and farming some of them may result in population explosions of others, shifting their status from minor irritants to major pests.

Beyond insect–soil, insect–microbe, and insect–insect relationships is the more familiar insect–plant buddy system.

In some cases, insects, like humans, will use plants for their own ends with no obvious advantage to the individual plants, but ultimately for what we might, in retrospect, characterize as the common good. In North America, for instance, about 1,700 insect species — mostly midges, flies, and wasps — are gall makers, commandeering plants' hormonal systems to create tumor-like growths, known as "galls," to provide homes and food for themselves. What's galling to the plants is that it decreases their seed production because it redirects their resources. The affected plants have obviously adapted; perhaps the galls limit plant growth and reproduction in such a way that they don't overpopulate and destroy their environments.

Mopane worms, the larvae of emperor moths (*Gonimbrasia belina*), are today considered a culinary delicacy in that cradle of human origins, southern Africa. What were they doing before our pre-human ancestors met them and invited them over for dinner? The mopane trees on which the larvae feed grow in woodlands across several southern African countries (Botswana, Zimbabwe, Namibia, northern South Africa). Elephants are the only other animals to eat a significant amount of mopane foliage; with their demise and retreat into parks, the caterpillars are a last defence against the slow, persistent march of dense, impassable, relatively infertile mopane-dominated veld across the landscape. Even where elephants are present, the caterpillars, which increase their body mass four thousand times during their six weeks of larval life, consume ten times more mopane than the elephants and drop almost four times more soil-enriching dung. In a word (okay, a few words), these larvae created, and still maintain, some of the most memorable and habitable landscapes in humanity's birthplace. Well, habitable for people and other charismatic megafauna, at least.

Palm weevil larvae are another group of bugs entering *haute cuisine* kitchens in the twenty-first century. Again, we would do well to understand the important work they were doing well before being discovered by previously non-insect-eating adventurers. Stressed, severely wounded, or dying palm trees speak to palm weevils (*Rhynchophorus ferrugineus*) in the language of volatile compounds (the arboreal version of perfume). The male beetles fly toward the source of the perfume, land on the suffering tree, and release their own pheromone messages to other males and females. The enticing chorus of tree and beetle perfumes brings more beetles, who then mate and lay eggs. The larvae, with sharp beak-like "noses," burrow into the palms and break down the wood; this is bad for the individual tree, but good for both the weevil and the ecosystem. Not only that, but because the weevil larvae are making nutrients available to trillions of other insects, fungi, and bacteria, and enriching the fragile tropical understory, they are providing a service to the forest of newborn baby trees. It was not until after postindustrial humans arrived and created plantations of date, oil, and coconut palms that the weevils were seen as pests. But is it their fault that we were not biologically mindful?

Orthoptera (grasshoppers, crickets, and locusts) are part of that web of herbivores that helped, and continue to help, recycle nutrients and determine the diverse, mosaic landscapes of natural systems that support other plants and other animals. They were doing excellent ecological work well before we reclassified them as plagues, protein, and, if you are poet Mary Oliver, beings to ponder and be amazed by, as in her most wonderful poem, "A Summer Day."

Like many Just Plain Folks, I tend to mix up grasshoppers and locusts, especially when they are on my plate. This is not surprising, I suppose, given how closely related the two are. However, locusts may provide only opportunistic disaster food relief whereas grasshoppers can become a staple, so we would do well to differentiate between them. In brief, all locusts are grasshoppers, but not all grasshoppers are locusts. Locusts are the ones that darken the heavens with their terrifying swarms; the shift from the solitary phase to the swarming phase is called phase polyphenism by the people who study orthopterans. If, like me, you prefer plain English, the locust transformation is a kind of Dr. Jekyll and Mr. Hyde story, or maybe an insect version of *The Incredible Hulk*, in which a regular, hard-working Joe is transformed into an uncontrollable, life-threatening beast. Of the more than ten thousand grasshopper species in the world, only about a dozen are classified as locusts. I'll talk more about locusts later, when I talk about pests, but for now, let's focus on the normal, Joe-Blow lives of grasshoppers.

In general, nonswarming grasshoppers are pretty much herbivores, eating a wide variety of plant species in a wide variety of rangeland and grassland habitats. In the past few decades, trying to get beyond the napalm/scattershot/neurotoxin–bombing approach to controlling them, researchers have looked more carefully at what these creatures actually do. Are they are always the villains we take them to be? Or are they a misunderstood spiritual totem? As one might have expected given nature's diversity, the answer is: both. Or, better yet: it depends.

Some grasshoppers, in some landscapes, eat plant species that, on their own, have slowly decomposing litter. This grasshopper feeding activity encourages plants that decompose more quickly, and this, in turn, speeds up nitrogen cycling and increases overall plant growth. Other grasshoppers, in other landscapes, prefer different plants and may suppress overall production.

Crickets too are rapidly becoming a staple of the entomophagy movement. Also belonging to the order Orthoptera, they are closely related to grasshoppers, locusts, and katydids. Like most of their relatives, they have taken to heart author

Michael Pollan's widely advertised advice to "Eat real food. Mostly plants. Not too much."[47] Among the two thousand or so cricket species in the world, most are herbivores who occasionally backslide into omnivorism. Like their grasshopper relatives in the wild, crickets are dedicated and diligent members of decomposer and nutrient-cycling communities, those unsung (but, in the case of crickets, singing) heroes of life on earth. They contribute by devouring huge amounts of cellulose-rich plant materials and producing frass (the technical name for insect waste products) that makes the energy and nutrients inherent in the plants available to bacteria, fungi, and, ultimately, to us.

Yellow mealworms, another entomophagical delicacy, are larval forms of *Tenebrio molitor* (darkling beetles), and members of a twenty thousand-species family — a family with dark secrets, judging from its Latin name, the Tenebrionidae. Some of us of a certain age are familiar with these animals as squiggly things in our mothers' flour bins. My mother sifted them out. Other mothers (perhaps Daniella Martin's?) may have fried them up and whisked them into an omelette. Other members of the same family are known as pests in grain-storage facilities and chicken barns, or as food for pet reptiles. Like other insects, different members of the darkling family have different jobs besides helping and pestering us. Various species within the family feed on such delights as decaying leaves, rotting wood, dead insects, dung, and fungi. In other words, they are more important and diligent recyclers than even the most obsessive blue box and green bin advocates. Furthermore, as part of the no-free-lunch ecological society, the great circle of life, and several other community-service clubs, they are themselves food for birds, small rodents, and reptiles, both in the wild and, more recently, those kept as pets.

Insects also played essential, nurturing roles in making sure the angiosperms —flowering plants — of the Cretaceous (100 million years ago), were properly cared for after the messy and violent divorce and fragmentation of the great Pangaea super-continent (which started about 200 million years ago). It is in the relationships between insect pollinators and flowering plants that we go beyond technical relationships and begin to understand just a little better the languages of conversation among insects and plants. In their roles as pollinators, insects and plants have evolved exquisitely intimate relationships.

The cacao tree (*Theobroma cacao*), a.k.a. the "food of the gods," originated millions of years ago in what is today Central and South America, although humans only began to make bitter chocolate beverages around 1900 BCE. One version of the story has it that the Aztec god Quetzalcoatl shared the secret of chocolate with humans, a transgression for which he was thrown out by the other gods. If one is inclined to make moral judgments, one might concede the rightness of their actions; apparently removing cacao beans from the pod is like removing a human heart during a sacrifice. Today cacao is grown in Africa and Asia as well as in the Americas, providing the cacao beans around which a US$50-billion industry is constructed. The tiny, white, downward-facing *Theobroma* flowers — which open at dawn and only last a day — grow directly off the lower branches of this tropical rainforest tree. The wild cacao flowers give off a complex perfume of more than seventy-five distinct aromas, which attract tiny midges of the families Ceratopogonidae and Cecidomyiidae — the only insects that can pollinate them. The midges, which do not travel far from home, require specific microhabitats in shady rainforests. In light of this information, growing *Theobroma* on small farms in the midst of rainforests, and protecting midge habitats, would seem to be a more intelligent strategy than clearing the forests for poorly producing plantations. This may seem unfair to the monocultural industrial cacao plantation owners, but then, I don't suppose the midges and flowers were thinking about treating corporations fairly when they eloped to the rainforest.

Fig trees offer an important illustration of how sustainable food systems, cultural identity, and economic class-based biases interact. These trees are important food sources for fruit bats, capuchin monkeys, langurs, mangabeys, Asian barbets, pigeons, bulbuls, and fig parrots, as well as the caterpillars of crow butterflies, plain tigers, giant swallowtails, brown awls, green garden loopers, and metalmark and "tropical fruitworm" moths. Fig trees also have symbolic importance in Hinduism, Islam, Jainism, and Buddhism (Buddha having found enlightenment while sitting under a fig tree). And remember Adam and Eve and the fig leaves? So, it makes sense that fig trees were a symbol of fertility in ancient Cyprus.

With a few exceptions, each of the thousand or so species of fig tree is (mostly) pollinated by its own companion wasp species. Not only that, but the female and male parts of the fig flower mature at different times. The wasp larvae grow inside the fig ovaries; the male wasp hatches and breaks into his sister's "bedroom," mates, and then (in remorse and anguish?) commits suicide. The now-pregnant female heads out, picking up pollen from the male parts of the flower, and flies to another fig tree where, pushing her way past the scales at the door, she pollinates the female parts of the next tree.

But not all figs and flowers are alike. In the 1880s, Californians imported Smyrna figs, which were known to produce delicious fruit. What the Americans did not realize was that Smyrna figs needed pollen from hermaphroditic caprifigs, which are apparently only good for goat food (a pretty low bar). Because of the way that Smyrna fig flowers are structured, fig wasps cannot reach in far enough to lay their eggs. The wasps lay their eggs in the caprifigs, where the young overwinter. In the spring, they fly out and pollinate nearby fig flowers, including those of the Smyrna figs, where they attempt, unsuccessfully, to lay their eggs. So the wasps need the caprifigs and the Smyrna figs need the wasps. In the 1880s, when American botanists first heard this story from European fig growers, who told them they needed to import both wasps and caprifigs in order to be able to produce figs for human consumption, they laughed. *LOL! Old men's tales! Idiot farmer hicks!* By the early twentieth century, still unable to get their fig trees to deliver the goods, they started importing.

Honey bees are highly valued in part because they are essential players in the maintenance of industrialized agriculture (almonds, canola, and cherries, for instance). They are also important for our sustenance in a variety of other, less obvious ways. In the 1990s, when I was involved in a research project on Brazil nuts, I discovered that before these nuts became fodder for cookies and muffins, they were part of a complex tale, one that involved bees, trees, birds, and rodents working together to bring them to fruition.

The Brazil nut tree (*Bertholletia excelsa*) grows in mature forests of the Amazon, only reaching reproductive age after thirty years, and can live as long as 1,600 years. During those years, the castaña, as it is called in Spanish, can grow to fifty meters tall, spreading its branches to thirty meters in diameter. Brazil nut forests only occur in the southwestern arc of the Amazon region, in an area that encompasses parts of Brazil, Peru, and Bolivia. The peak flowering of the castaña is in October, November, and December. During this time, in a daily cycle, the trees produce large flowers that fall to the forest floor.

The crown of the castaña provides a space for wild *Stanhopea* and *Catasetum* orchids, which, like all orchids, are epiphytes (plants that grow harmlessly on other plants). They need a place to hang out, gathering water and nutrients from the air and the surface of the tree. These orchids attract the orchid bees, especially the romantic vagabond males looking for the right cologne and the right female. Male orchid bees, with their large bodies and big tongues, are one of the few insects capable of penetrating the large and heavy outer petals of the Brazil nut flower to pollinate it. So, while they are hanging around the pretty orchids, pollinating them and picking up a female-attractant, they also visit the castaña flowers.

As the Brazil nut pod matures on the tree, macaws feed on the fruit, thereby lightening the load of the branches and protecting them from breaking due to excessive weight. The tough-skinned two-kilogram pod, when ripe, crashes to the ground, where it is collected by the brown agouti (*Dasyprocta variegate*), a large rodent that is the Brazil nut tree's main seed disperser. The agouti cuts through the pod's hard shell to get to the nuts, and those it does not eat, it buries for later consumption. Of course, the agouti doesn't always remember where he left his nuts. Those are the ones that, over decades, can replenish the forest. The next time you crunch into a cookie with Brazil nuts, think about bees, orchids, and the rain forest.

Among the lessons entomophagists will need to heed again and again as insect-eating expands geographically are that the insects we care about (whether as plagues or as food, or, in the case of butterflies, for their beauty) may be vulnerable in unexpected ways. Thinking laterally — as from Brazil nuts to orchids to bees — is essential if we are to even begin to understand the ways in which humans and the millions of mostly unknown insects influence each other.

The Brazil nut case is but one illustration of how little we understand the "languages" used by insects, which we will need to learn if we are to more deftly manage our entomophagical relationships with them. For other livestock, understanding the meaning of their behaviors, the sounds they make, and the chemical signals they give off have had important implications for breeding, selecting for desirable traits, and diagnosing illnesses. How, then, do insects communicate?

Many insect predators, including people, are familiar with the stinging chemical warning shouts and bites of hornets, bees, and ants. Entomologist Justin Schmidt, sometimes called the "Connoisseur of Pain," has documented and rated many of these using a pain-measurement scale. To focus on these toxic stinging and biting forms of cross-species communication, however, would be like lumping the spears and gunshots fired by people — which of course bear their own meanings — together with the complex subtleties of spoken and body languages. One of our late learnings with regard to managing cattle is the relationship between hormonal cycles, milk production, reproduction, vocalization, and behavior around other cattle. For insects, the counterpart to hormones is pheromones.

Jean-Henri Fabre was a nineteenth-century French naturalist and the author of ten volumes of *Souvenirs entomologiques*. Derided by "serious" scientists because he wrote so that non-scientists could understand, today he is best known, and widely celebrated, in Japan. An anti-evolutionist, he was nevertheless praised by Darwin for his careful and keen observations, particularly on insects. In 1874, Fabre noted that a newly emerged female peacock moth, stretching her wings under a wire-gauze bell jar, was attracting males from very far away. This observation led to the identification of so-called "calling glands" in insects. Over the next century, the chemicals released from these glands came to be called pheromones. One of these perfumes, released by the female Indian luna moth, could be detected by males over ten kilometers away. Insects release pheromones that are like specialized dating sites, enabling them to find members of their own species. Some are designed for long-distance seduction. Others, released by males, act over shorter distances and are termed aphrodisiacs, which May Berenbaum describes as "excitants — something to put the female in the mood."[48] There are even "hey boy, back off!" pheromones, such as the "mace" squirted at persistent males by an already-mated female *Pterostichus lucublandus*. There are alarm pheromones, pheromones that synchronize group activities, those that protect areas where eggs have been laid, and those that serve as trails to food supplies.

We have already encountered the link between the scents of orchids hanging out on Brazil nut trees and the bees that pollinate the trees. Orchids might be seductively pretty, but they are also devious. The orchid *Dendrobium sinense*, endemic to the Chinese island Hainan, is pollinated by the hornet *Vespa bicolor*. The flowers' challenging problem is that the hornets prefer to catch honey bees and feed them to their babies; they're not much interested in pretty flowers. So the orchids produce a chemical component of the alarm pheromones of both Asian (*Apis cerana*) and European (*Apis mellifera*) honey bees. The scent of the flower, mixed with the alarm pheromone, attracts the hornets, who pollinate the flowers while hunting the bees.

This brings to (my) mind the beekeeper anecdotes about bananas and bees. The scent of bananas also mimics the alarm pheromones of honey bees. Since banana flowers in the wild are pollinated by bats and birds, is this scent a way of bringing in bees so that the pollinators are attracted by the scent of a bee lunch?

I am reminded of the time I accompanied my son Matthew, who owns and operates Unspun Honey in Mt. Gambier, Australia,[49] when he was called to remove a swarm that had reportedly settled into a black commercial compost barrel in someone's backyard. A swarm, which is a homeless colony of bees with a queen scouting for a new place to live, is generally not aggressive. They don't have a home to protect and are busy negotiating with scouts about new possibilities. We carefully approached the buzzing composter across the small backyard, stepping over children's toys. Matt lifted the lid and looked in. This was not a swarm. These bees had already decided that the composter was a cozy suburban home — protected from the weather, near flowers — and had started to build comb on the underside of the lid. When Matt began to scoop the bees into his pail, they became irate and began to attack him. He went back to the truck and returned with a hive box, into which he had put a "lure" — lemongrass, which mimics a bee pheromone that says "this is a good home." He dumped in as many bees as he could, including the queen, and then left the box there, propped on the edge of the composter bin, for the day.

When he returned that evening, much to his and the homeowner's relief, the bees had all settled into their new, lemongrass-scented home, and the compost bin was empty. While orchids and banana plants may be able to mimic the bee alarm pheromone, lemongrass makes them feel at home.

Pheromones provide potent ways for insects to communicate and are therefore important to consider as we develop insect management and farming practices. But pheromones are not the only insect languages. Like people, insects also learn about their surroundings through visual and aural cues and send messages through the sounds they make. The nature of these messages among insects, however — how they are sent and received — often differs qualitatively from what we have come to consider normal through observing other species, including our own.

"Listen," said Matt to me as we paused amid the sunny buzz of bees whizzing over our shoulders, to and from the hives. "You can *hear* that the hive is happy." I listened, and thought of Mark Winston's assertion in his 2014 book *Bee Time*, that "a tacit understanding exists between beekeeper and colony; if you're calm around honey bees, they will be calm as well, creating a dynamic that feels to the beekeeper like a relationship."[50] When I first started thinking about insects and sounds, my thoughts went to the soothing or angry sounds of Matthew's bees, and then to chirps and trills of male breeding-age bush crickets (katydids) and to the cheerful chirping of eager crickets that I'd heard in small household breeding pens in rural Lao PDR and the warehouse-sized Entomo barn in Canada. The male crickets chirp that they are ready now; where are the females? We haven't much time! For the forager or farmer of insects, these chants, trills, and songs are important signals, messages from the insects to those who aspire to manage them.

Non-insect-eating urbanites, especially those of us who grew up listening to rock 'n' roll and working in factories without safety protection earmuffs, are most attuned to very loud sounds. The loudest sound produced by any animal on the planet, on a body-weight basis, comes from the tiny lesser water boatman (*Micronecta scholtzi*), who can generate ninety decibels by rubbing his penis against his abdomen. Eggs from another member of this family, the axayacatl water bug, were sold by the Aztecs; the Spaniards called it Mexican caviar.

While I have not heard the water bugs or tasted their caviar, I have had my ears punished by the heavy-metal shrieking of the East Coast brood of the genus *Magicicada*. These animals, celebrated, eaten at opportunistically celebratory banquets, and despised as a bug plague, emerge dramatically and with predictable periodicity from the soil. The broods, which are staggered so they don't all come out dancing across the United States at the same time, are labeled by Roman numerals. For these periodical cicadas, found only in North America, the number of years between one emergence and the next is always one of two prime numbers — thirteen or seventeen — a phenomenon that no one has yet explained in any convincing manner. Just before the big bio-dance, the nymphs burrow up to just below the surface, ready to enter the stage all at once. When the temperature and moisture are just right, they emerge by the millions and millions, crawl up the nearby trees, shed their skins, and have a coming out party. For any animal that eats bugs — fish, small mammals, turtles, birds, and people — the cicada emergence is a once-in-a-lifetime belly-stuffing bonanza.

The soft, white adults harden and darken within hours and then fly clumsily about, the males singing boisterously. This time in the lives of the *Magicicada* is described by musician-philosopher-naturalist David Rothenberg (in his book *Bug Music*) as a "few weeks [of] nothing but revelry, music, and sex."[51] About ten days later, they mate, and then, as seems to happen after all such parties, the fun is over. The females deposit about 500 eggs each in slits they make in the tree bark. Six or seven weeks later, by which time the adults are dead, the little nymphs hatch, drop to the ground, and dig in for some prime years of resting, nibbling roots, and sucking phloem. When the adults die, their bodies become a huge dose of fertilizer for the elm, maple, oak, and ash trees they prefer. Quite apart from the noise they make, the periodical cicadas are important long-term recyclers of nutrients in the deciduous forests.

Much of the ecological soundscape of insects is, however, much quieter, not intended for human ears — especially not for those of us with eardrums damaged by rock 'n' roll and industrial work.

In 1977, Canadian composer, teacher, and musician R. Murray Schafer published *The Tuning of the World*, in which he invented what has come to be called — in research journals and books — acoustic ecology. Schafer has argued that we pay too much attention to the loud noises, the birds and crickets; as a result, we miss the complex, subtle communication systems among a variety of insects and plants. In 1992, musician and composer David Dunn released *Chaos and the Emergent Mind of the Pond*, a recording (and rearranging) of the rhythmic clicks and pops and buzzes of aquatic insects in North American and African ponds. Almost fifteen years later, he released *The Sound of Light in Trees: The Acoustic Ecology of Pinyon Pines*. Using tiny microphones and transducers placed in the phloem and cambium of pinion trees of the southwestern United States (*Pinus edulis*), he recorded the "voices" of, and conversations among, pinion engraver beetles (*Ips confusus*) and, possibly, bark beetles (*Dendroctonus*), as well as larvae of various wood-boring and longhorn beetles. He and his colleagues compared infested trees with healthy trees and proposed that what they called "bioacoustic interactions between insects and trees" were "key drivers of infestation population dynamics and the resulting wide-scale deforestation."[52]

In 2015, a team of Spanish researchers reported on a study they had done on two tiny mirid bugs, *Macrolophus pygmaeus* and *Macrolophus costalis*, which feed on aphids, whiteflies, and other pests of vegetable crops. Using "laser vibrometry" (which is not, apparently, a sex toy), they measured what they called "substrate borne" vibrational signals. What they discovered was that these minute bugs used communications built around two distinct sound types, each with its own harmonic structure. The first, a "yelp" sound, is the basis for an important pre-mating song. The other seemed to be associated with an increase in the time the bugs spent walking.

With the exception of blind people, most of us rely on sight to orient ourselves in space, and we often assume that our visual world is similar to those of insects. We know, however, that bees are "red-blind." Like humans, bees have three photosensitive pigments, but they encompass the ultraviolet blue-green range; that is, they are sensitive to ultraviolet wavelengths, below 380 nanometers, which we cannot see. We have blue-green-red receptors, but the bees see only black where we see red. Dragonflies and butterflies, on the other hand, may be pentachromatic. It is difficult to imagine what the world looks like if one has five built-in color channels.

Of course, vision is not just about distinguishing colors. Flowers that are viewed through an ultraviolet filter such as those of the bees probably look quite different than they do in what we consider "normal" light; some, like black-eyed Susans, apparently have a bull's-eye pattern to attract pollinators; others have what looks like a runway to guide insects into landing. Further distancing their visual worlds from ours is the fact that many insects that go about their lives in daylight have compound eyes with multiple, independent light-gathering units, called ommatidia. Inputs from the ommatidia are brought together in the brain, where a single upright image is created. This is called an apposition eye.

Fireflies, moths, and many insects that fly at dusk or in the dark have what are called superposition eyes, which are 100 times more light-sensitive than the appositional eyes of diurnal insects. In superposition eyes, the ommatidia cooperate and only project one image back onto the retina. When our family lived in Java, I would sometimes go out into the darkened countryside to look up at the stars in the clear heavens, unpolluted by the haze of city lights; looking across the dark, flooded rice paddy, I was awed by the shimmering sheets of fireflies (actually beetles of the Lampyridae family), their lights flashing on and off in synchronous waves. What I did not understand at the time was that I was seeing mate-seeking male fireflies flashing species-specific light messages; females of the same species would respond with their own unique light, sending a message: "Here I am, ready, willing, and just waiting for you!"

Being able to send, interpret, and receive appropriate light signals among a cacophony of signals from similar species is, for these beetles, a matter of survival. Unfortunately for *Photinus* males, the related but different *Photuris* females have figured out the code of several other species, including *Photinus*. When the *Photinus* males respond eagerly to the come-hither signals, the *Photuris* females eat them. I guess it's one way to eliminate the competition and have a good meal at the same time.

Neither the apposition bees' eyes nor the superposition fireflies' eyes have the ability to focus, nor can they move in their sockets. They are, however, much quicker and more efficient at detecting motion than our eyes are, which is why insects are so difficult to catch, and which has implications for those who wish to create foraging techniques that don't involve excessive collateral damage.

Ball-rolling dung beetles, which are eaten in many parts of the world, use patterns of light polarization and visual cues from sunlight, stars, and, more specifically, constellations such as the Milky Way. Diurnal beetles, which are active during the day, have compass neurons that respond to sunlight. The neurons in the brains of nocturnal beetles can also respond to polarized light from the moon, and light cues that are a million times dimmer than those that guide the day-shift workers. The beetles need to move in a straight line away from the dung pile in order to avoid competitors and thieves; before they start moving, they climb up on the dung ball and, dancing with the stars, orient themselves. Once they figure out where they are, and where they need to go, they begin to roll the ball, occasionally climbing up to dance again if they lose their way.

Beyond sight, sound, taste, and touch, some insects use magnetoreception, which is the ability to use magnetic fields for orientation in the landscape and navigation to new areas. This ability is well documented in ants, termites, and honey bees.

For insects, as for people, sensing the world is necessary but insufficient. How those sensory inputs are integrated and used determines sustenance and survival. Insects live in a world with a very different sense of space and time than our own. Even if we set aside the senses we can describe but hardly imagine, such as those related to magnetism or the response to gradations of polarized light, the colors and motions are only a small part of the picture. As your friendly neighborhood neuroscientist can tell you, seeing something is not the same as perceiving it. What, exactly, is it that we're seeing? The process of perception involves integrating and assessing multiple sources of information to make decisions: is this food? An enemy? Not relevant? Bees have evolved ways of integrating all these perceptual sources and have used them to create a language that rivals ours in complexity. They integrate complex dance moves with scents and orientations to the sun to communicate among themselves, describe location and quality of nectar and pollen sources, and negotiate among several possible new homes for a swarm.

In insects, we attribute process to some kind of neuro-physiological or neuroanatomical algorithm. In people, the same kind of data are used to demonstrate abstract reasoning and are evidence that we have minds.

What should be clear by now is that insects live in a world we struggle to understand. Just as the entanglements of our social world are a legacy of the stories and battles among our proto-Indo-European, Afro-Asiatic, Dravidian, and Sino-Tibetan human ancestors, so the complex natural world we inhabit is the cooperative legacy of conversations among the "millions and millions" of worldviews and languages of insects. They conceive of the world in terms of color (within a range different from ours), sound (within a range different from ours), scent (within a range different from ours), and, more closely related to human senses, taste (sweetness, sourness, saltiness, bitterness, and umami) and touch (pressure, pain, temperature). Insects have all these senses and more, including some, like magnetoreception, which we can scarcely imagine. They experience and understand the world in different ways — qualitatively and profoundly different ways — than humans.

Hugh Raffles, summarizing the work of twentieth-century biologist and philosopher Jakob von Uexküll, says that "All beings live in their own time worlds and space worlds . . . distinct worlds in which both time and space are subjectively experienced through sense organs that differ radically among beings and produce radically different experiences."[53]

If, in the new green world of entomophagy, we aspire to avoid the excesses of industrialized livestock agriculture and the ecological impacts of overforaging, we will need to pay attention to the languages and conversations among insects and their worlds. At a technical level, these conversations are important for the management of pests and grub. For me, this goes beyond the merely technical. As we think about bringing insects more fully into that circle of ecological intimacy we call eating, I am in awe of these insect multiverses.

## PART III. I ONCE HAD A BUG: HOW PEOPLE CREATED INSECTS

Insects and people have been warring against each other for aeons. Our stories have prepared us to be on alert, battle-ready, when we see them, or, failing that, to run for our lives. How do insects fit into our cultural imagination? Have we cast them all as monsters and aliens so that we can more easily, and without compunction, kill them? Let us explore the dark narratives that have driven our agriculture, food, and disease control policies and have brought us into the battle-scarred landscapes of the twenty-first century.

I'M CHEWING THROUGH YOU

*Insects as Destroyers and Monsters*

• • • • • • •

*I once had a bug, or should I say,*
*she once had me?*

The conflicting ways in which humans have imagined the insect world have led to equally conflicting responses to it. Until the late twentieth century, the dominant global narrative, driven by a narrow European view of science and technology, was of insects as evil — pests to be killed in a take-no-prisoners war. For those who continue to think this way, "eating bugs" is viewed as disgusting at worst, and problematic at best — since it is not usually considered good manners to spray insects with poisons and then offer them as food.

For most of us, insects are strange, not human, *other*; uncontrollable, reproducing by the millions, they can invade our homes and even our bodies, bringing disease and destruction, evading us with their deft skittering and slithering movements. And then when we see them on the plate, how quickly, unconsciously, and irrationally these fears and anxieties, borne of a history of pandemics and pests, can morph into the more visceral feeling of disgust. For the new entomophagists, this is where the biological rubber hits the cultural road, where gut-level personal experience tangles with abstract information. We are all into being green and ecologically friendly. We all want to feed the world sustainably. But with *worms* and *grasshoppers*? Really?

Imagine this: there is a terrible churning in one's abdomen, then pressure as the belly distends, and then extreme pain as the muscles are ripped open and a newborn animal appears. If, as in the *Alien* movie franchise, that newborn is insect-like, we are, even beyond imagining the extreme discomfort of a large animal bursting out from one's abdomen, horrified, disgusted, even fearful. A wide-eyed loris emerging under similar conditions might evoke more mixed feelings — *OMG how cute!* — alongside the discomfort. And if the emergent life looked like E.T. from the eponymous movie, we would be certain that we had entered the realm of goofy farce.

In non-insect-eating cultures, the ones that through violence and commerce have sought to impose their ways of life on the planet, large-scale insectiform beings are invariably used in film and literature to evoke horror.[14]

If J.B.S. Haldane's scientific mind saw the Creator as having had an inordinate fondness for beetles, then the vivid insect scenes in novelist Malcolm Lowry's dark, disorienting tale of an alcoholic's tragic descent into hell, *Under the Volcano*, would seem to invoke the Hindu creator god's other half, Kali the Destructor. Lowry's Consul watches "helplessly" as the "mosquito-stained" walls around him swarm with insects, as though "the whole insect world had somehow moved nearer and now was closing, rushing in upon him." According to Spanish film director Luis Buñuel, Salvador Dali suffered delusions of parasitosis that led to self-mutilation. From *Men in Black* and *Aliens* to Gregor Samsa's transformation into a cockroach-like animal in Kafka's *Metamorphosis*, insect-like invaders and transformations are, in non-bug-eating cultures, invariably bad news.

Large groups of locusts, ants, or beetles, unlike large groups of cattle, do not bring to mind pastoral scenes from the Great Plains or biblical references to cattle on a thousand hills. Large groups of insects swarm. They invade. They infest. As long as they respectfully keep their place in nature, some insects might be revered, as were the Egyptian scarabs, or held up as paragons of virtues such as hard work, as with the ants in Aesop's fables or the proverbs of Solomon and honey bees in Christendom. Most often, though, bugs send shivers down our spines.

The stories that have created this unsettling quiver come at us from all angles and reflect the complexity of nature and our multiple roles in it. More information merely compounds the problem of what we should do. For every story of ants cleaning garbage off the streets of New York, there are stories of armies of Argentinian fire ants eating baby caimans. The implied, unstated question is: would they do that to us? And who among us could forget the swarming red ants in Gabriel García Márquez's *One Hundred Years of Solitude*, the scene in which "all the ants in the world" are dragging the dead baby to their holes?

Nature writer David Quammen argues that our desire to see insects as horrible is deeply rooted in the human psyche. His book *Monster of God: The Man-Eating Predator in the Jungles of History and the Mind* is a moving lament for the terrible predators that have throughout history stalked and devoured us like just another fresh-meat meal. These animals — the lions and tigers, crocodiles, bears and wolves — still occasionally steal children and strike fear into isolated populations living near wilderness areas around the world. Even as we drive them to extinction, they remain within us, lurking in the dark jungles of our psychological ecology. In the last chapter of the book, Quammen reflects on the insect-like Alien created by the late H.R. Giger for the eponymous movie: "I believe that the success of the Alien series, like the durability of *Beowulf* and *Gilgamesh*, reflects not just our fear of homicidal monsters but also our need and desire for them. Such creatures enliven our fondest nightmares. They thrill us horribly. They challenge us to transcendent fits of courage. ... The only thing more dreadful than arriving on LV-426 and finding a nest of Aliens, I suspect, would be to arrive there, and on the next unexplored planet, and on the next after that, and find nothing."[55]

Some think that science, by demystifying nature, can save us from this need for monsters, that in the face of our incredible technology and domination of the planet, we will see these monsters as figments of our feverish imaginations. But the human mind is not so easily compartmentalized, and the roots of the non-insect-eater's revulsion for insects has its roots in science as well as the imagination. Scientific reports and nightmarish monsters reinforce each other.

Stepping outside the movie theatre where we have seen the imaginary Alien, for instance, we encounter insects that eat the flesh of living animals. Miasis is a condition in which flies lay eggs on the skin of animals, and the maggots feed on the living flesh. Flies that engage in this unwelcome behavior are called botflies or blowflies. These flies are tiny, but what if (our movie-fed imaginations now grown hyperactive) they were huge?

*Bot* derives from a Gaelic word for maggot, and, according to the late British forensic entomologist Zakaria Erzinçlioglu, the word *blow* historically meant a mass of fly eggs. Hundreds of years BCE, Homer wrote about "the blows of flies," and in the early 1600s Shakespeare wrote in *Love's Labour's Lost* that "These summer flies have blown me full of maggot ostentation" and, in *Antony and Cleopatra*, "Lay me stark naked and let the water flies / Blow me into abhorring." This gives the more recent expression "blow job" a somewhat different slant than what adolescents think it means. Thankfully, encounters with fly maggots that eat living human flesh are a rare occurrence. More commonly — and in many ways more devastatingly — we are surrounded by those that drink our blood.

In 2007, I was part of a mission sponsored by the World Organisation for Animal Health, also known as the OIE.[56] Three of us — all of European descent — were to assess the animal health capacities in Cambodia. In less than a week, we drove between flat fields of rice in the south, pondered the flocks of ducks bobbing around in rickety enclosures that extended from the fields into one of the many tributaries of the Mekong, and nosed through offices from the Vietnamese–Cambodian border in the south to the north end of Tonlé Sap, a thriving lake that occupies much of the center of the country. From pre-dawn to after dark, we visited Spartan government and lush private laboratories, hole-in-the-wall pharmacies, makeshift autopsy rooms, research centers, colleges and universities still struggling to reinvent themselves after the Pol Pot regime all but destroyed them with its brutal anti-intellectualism, outdoor slaughtering "slabs" where Brahman cattle could die breathing the sun-filled air, and half-hidden slaughterhouses where little boys in torn, bloodied shorts scrambled with plastic buckets after gushes of pigs' blood, gut-splatters, and bits of discarded flesh. We saw chickens, ducks, pigs, and cattle, free-range and caged, from scruffy to fat to dead and hung up on hooks. On the road trip between Phnom Penh and Siem Reap I saw something else — and yet, in some Douglas Adams, if-it-doesn't-make-sense-it's-invisible, sort of way, I didn't see it.

Scattered across the green rice paddies and along gray, water-filled ditches and ponds were structures from which flapped sheets of translucent plastic. Below each sail was a rectangular "boat" constructed of the same materials. In the daylight they were usually rolled up, but at night the sails were unfurled into vertical rectangles. More intriguing to me, they were lit up at night, like ghostly square-faced scarecrows rattling and flapping in the wind and rain. When I asked about these, our guide smiled. At the next roadside market, he showed me large baskets heaped full of walnut-sized water bugs. The beetles, attracted to the light, flew into the sheets, and then dropped into the containers below. They were food. For people.

At the time, I thought this a slightly unsettling curiosity and recalled how, in 1968, lugging a heavy, frameless, green canvas backpack through the same region, the back of my shirt so soaked with sweat it would later peel off in shredded pieces, I had declined to nosh on the beetle "shish ke-bugs" proffered up to my open window by eager boys at Thai bus stops. Now, again, I found the very idea of biting into these cockroach-like bugs revolting.

I did not know, in 1968, or in 2007, that giant water bugs were grilled or fried in Thailand, Lao PDR, and Cambodia, or that the Thais ate so many that they were importing them from neighboring countries, that wild populations were declining because of habitat change and pollution, that prices were going up and that they were difficult to farm because the bugs started eating each other when they were crowded. It had never crossed my mind that caring for these bugs might be part of my veterinary job, or that the OIE and the Cambodian Ministry of Agriculture should be interested in this. Or even that these insects might hold one of the diverse keys to local food security around the world.

Why did these things never cross my mind? I suspect it was because I was distracted by an immediate concern for the well-being of my own species.

What I did see were the effects of people being food for insects. I saw a motorbike, man in front, woman behind, infant in between. This is not an unusual sight in Southeast Asia. Often you can see two or three children along with their parents and full shopping bags on a tiny motorbike. The difference here, in Phnom Penh, in 2007, was that the woman was holding up a bag of intravenous fluid, and the thin tube from that bag was inserted into the arm of the infant. Cambodia was in the throes of an epidemic of hemorrhagic dengue fever that, in 2007, sickened 40,000 people and killed more than 400, many of them children. Dengue fever viruses — like those that cause Zika and yellow fever — are spread by mosquitoes. The female mosquitoes are the ones that feed on mammalian blood, and hence the ones that transfer the virus from one bloody lunch buffet to the next. Ironically (if one can see past the obvious species barriers), the female mosquitoes are drinking human blood in order to ensure that their own offspring live. Mosquitoes are manifesting that fierce mothering instinct that has enabled many species to survive these many centuries in the face of the incalculable odds against them. They are not "out to get us." Like the Maasai who drink the blood of cattle, and the French who eat blood sausage, the mosquitoes, lice, and ticks that use our blood are simply nourishing themselves and their babies.

Mosquitoes are not the only blood-feeders. Triatomine (a.k.a. assassin) bugs are one of about seven thousand species in the Reduviidae, a Hemipteran (true bug) family of predators and bloodsuckers. Although there are reports of some Aboriginal people eating Reduviidae in central Australia, we have more often thought of these bloodsuckers as eating us. The triatomines feed on people and other animals, having probably evolved from insect-eating ancestors more than 175 million years ago. Triatomines can be infected with *Trypanosoma cruzi*, a wavy, willow leaf–like flagellated hemoparasite. The trypanosomes, which cause various forms of mammalian and marsupial sleeping sickness in Africa, Australia, and South America, appear to have evolved from parasites that first made their homes in insects. Although the great continent of Pangaea started breaking up about 200 million years ago, it wasn't until a few tens of millions of years later that Africa and South America went their separate ways and the parasites followed, wagging their tails behind them.

Emerging at night from dark crevices in mud walls, these triatomines wander down walls and hammock ropes. Finding a sleeping person, they look for a good place to get some blood, usually an area of exposed mucous membrane such as next to the eye or lips. This gives them their other popular name, kissing bugs. So as not to disturb the person's sleep, they start by injecting a little anaesthetic. Then they suck blood from, say, the medial canthus of the eye, take a crap, and wander back home. The person wakes up, rubs his itchy eye, and, in so doing, rubs the parasite (which is in the poop) into it.

A few people get an allergic reaction at the site of the bite. Most people carry the parasite without getting sick. About 10 percent develop a chronic disease that can result in a weak and flabby heart and/or a dilated, flabby intestine and esophagus. These symptoms may take decades to develop. As you can imagine, folks with the chronic disease lack energy and just generally feel miserable.

Even Darwin, in the midst of his celebration of science and "objective" observation, could not suppress or disguise his disgust at bloodsucking insects. In *The Voyage of the Beagle*, he wrote: "We crossed the Luxan, which is a river of considerable size, though its course towards the sea-coast is very imperfectly known. It is even doubtful whether, in passing over the plains, it is evaporated, or whether it forms a tributary of the Sauce or Colorado. We slept in the village, which is a small place surrounded by gardens, and forms the most southern part, that is cultivated, of the province of Mendoza; it is five leagues south of the capital. At night I experienced an attack (for it deserves no less a name) of the *Benchuca* (a species of Reduvius) the great black bug of the Pampas. It is most disgusting to feel soft wingless insects, about an inch long, crawling over one's body. Before sucking they are quite thin, but afterwards they become round and bloated with blood, and in this state are easily crushed."

In the years following the voyage of the Beagle, Darwin suffered from lassitude, palpitations, extreme fatigue, flatulence, and general gut problems. At least forty different diagnoses have been proposed (ah, to be an armchair physician!). Among those, several authors speculated that he had picked up trypanosomiasis, and Zakaria Erzinçlioglu has even wondered if Darwin would have written *The Origin of the Species* if he hadn't been chronically sick, having to stay home to write instead of being out and about with his naturalist pals. Having once suggested that Gaudí might not have designed the Sagrada Família if he hadn't suffered from chronic brucellosis acquired by drinking goats' milk (as a natural health food), I have some warm feelings, but no evidence, for this theory. When I am sick at home, I don't do much writing. But maybe that's just me.

Other researchers suggested that it was a mutation in his mitochondria that made Darwin sick, which would make it his mother's fault, since mitochondria are inherited through maternal lines. Of course! It's always the mother's fault!

Not all of our battles with insects are the consequences of direct attacks on human populations. In many cases, our anger, fear, and disgust are the result of attacks on foods that we love. Among the mythologies that have shaped European, African, and American relationships with the natural world, few are more powerful than the tales of the locust plagues visited upon Egypt. The plague described in that founding myth for the Jewish people, gathered from oral traditions during their Babylonian exile in the sixth century BCE, has retained its power because of the frequency with which humanity has reexperienced and recounted its devastation.

Locusts are the Darth Vaders of the insect world, the good guys gone bad. Unlike the majority of grasshoppers, locusts have two phases, one suited for a solitary, albeit crowded, life, and one designed for swarming. During years of high productivity, locusts in the solitary phase reproduce even better than rabbits. The population explosion that results, aggravated by droughts and declining food supplies, causes chemicals to be released in their frass and increased disturbance of their leg hairs as they crowd. Under these conditions, females lay eggs that are biochemically primed to grow up into nymphs that, under endocrine changes stimulated by continuous crowding, grow longer wings and actually seek to aggregate. Like the European explorers of the Age of Empires, and the refugees from overcrowded hovels in medieval Europe, they are anxious, pumped, and primed to swarm. Better to leave home, the endocrines say, than to stay and starve. During a swarm, locusts abandon their vegan lifestyles in favor of voracious, omnivorous feeding frenzies.

In the nineteenth century, immense and catastrophic locust plagues seemed to come out of nowhere, in no predictable pattern, roiling in dark feeding frenzies across the American Midwest. Politicians, religious leaders, and scientists described these plagues in language that resonated with the weight of religious tradition and the fearful fury of the mysterious wilderness. Then, suddenly, at the end of the century, they disappeared. No more plagues.

Would they ever come back? Without knowing where the swarms had come from, where they lived between plagues, and why they had disappeared, Midwestern American farmers and settlers might forever be scanning the horizon for another approaching catastrophe.

Before the work of Jeffrey Lockwood and his colleagues in the 1990s, entomologists had been searching for causes of locust extinction commensurate with the size of the plagues: massive changes in landscape, disappearance of the bison, widespread planting of alfalfa, climate change. One day, after years of painstaking and fruitless research, Lockwood was talking with a colleague about monarch butterflies. Monarchs migrate thousands of miles down the length of North America to overwinter in one small stand of trees in Mexico; the adults feed only on milkweed. They are vulnerable at various stages of their complicated lives, and much has been made of reducing pesticide use and promoting milkweed growth on their flight paths. This is all well and good, but monarchs are probably most vulnerable to extinction while in their tiny Mexican sanctuary. That's where the babies are born. That's their only nursery. If those trees go, then the monarchs disappear, regardless of reduced pesticide use and butterfly gardens. Lockwood started wondering: what if swarming locusts were, like monarchs, reliant on some small, unknown sanctuary? He decided to pursue this possibility further.

In his entomological thriller *Locust: The Devastating Rise and Mysterious Disappearance of the Insect That Shaped the American Frontier*, Lockwood reported that the locusts had retreated, century after century, to river basins in the northern Rocky Mountains to regroup. Then, the fur trade took away the beavers, and the locust sanctuaries were increasingly subjected to spring flooding. In the late 1800s, prospectors and miners headed into those mountains after gold and silver. The miners needed food, and farmers followed them into the fertile valleys with sheep, cattle, and alfalfa crops. The disappearance and extinction of the plague locusts was an unintended consequence of a combination of land-clearing and intensive farming practices in support of mining communities. More to the point, it wasn't simply the result of changes all across their range, but also — probably mostly — because of a direct but inadvertent attack on their nurseries.

While many of us have grown up with some awareness of locust plague stories, few of us are aware of the many insect plagues that attack the foods we love. The plant pest grape phylloxera (*Daktulosphaira vitifoliae*) is a prime example of such a pestilential attack.

In his 2005 *New York Times* review of Christy Campbell's compelling tale *The Botanist and the Vintner: How Wine Was Saved for the World*, reviewer William Grimes declared: "What the Black Death was to humans, the phylloxera epidemic was to grapevines: a mysterious, unstoppable killer that ravaged not only France but nearly all the wine-growing world."[57]

The nineteenth century was an exciting time for European and American explorers and natural scientists, people who actually went out into the world and paid attention. These were the heady days when the amateur naturalists Charles Darwin and Alfred Russel Wallace were pulling together disparate travel tales about their observations of birds, barnacles, and beetles. The microbe hunters Robert Koch and Louis Pasteur probed the Lilliputian world of bacteria, yeasts, and fungi. Even as these heroes of modern science and medicine were reimagining the world we lived in, entomologists, botanists, and entrepreneurs, driven by a desire to make profits as well as by public concerns about locust plagues and pest outbreaks in agriculture, were digging more deeply into the world of applied, problem-solving sciences related to agriculture. These were years when no one thought much about killing specimens to study them, or carrying plants and animals back and forth across oceans to stimulate and improve agricultural and food supplies. Some of the plants that made their way across the Atlantic from America to France were grapevines. Although the vines themselves seemed hardy, disease resistant, and productive, the wine produced was considered repugnant. Nobody, at first, noticed the tiny aphid stowaways that arrived in the mid-1850s.

Over the decades from the 1850s to the 1870s, however, wine growers in France could not help but notice the shrivelled leaves and blackened, rotten roots of their grapevines. Businesses were going belly-up. Wine was big business in France, but politicians were busy liberalizing the economy, and traditional *viticulteurs* were distressed and bewildered. Between 1875 and 1889, annual wine production plunged from 84.5 million hectoliters to 23.4 million hectoliters. Two-thirds to nine-tenths of all European vineyards were destroyed. Intensive investigative work by French botanist Jules Émile Planchon and American entomologist Charles Valentine Riley identified phylloxera as the cause of the blight, and they began piecing together its complex life cycle. But what to do about it? Desperate interventions, like burying a live toad under each vine, did not seem to help. After much grumbling, French wine growers went along with a suggestion from two of their own that they could graft European vines onto American roots. This maintained the European quality of the wine, but took advantage of the disease resistance of the American roots. Gradually, the French wine industry recovered. Nevertheless, the political, cultural, and scientific squabbles continued. In California, phylloxera made a resurgence, perhaps evolving (as all insects will) to adapt to the American roots. Some vineyards in Europe had not succumbed to the blight, and those were nurtured and studied for new answers. Even the once reviled American hybrids have been resurrected in France. According to Campbell, Mémoire de la Vigne is an association that commemorates the times when the only wine available to the peasants was American. Some have declared this to be "the wine of resistance, the wine of the anarchists, the wine that drives you mad," which, translated into Canadian English, means (I think) that they liked it.

Insects have thus injected us with parasites, chewed their way into our skin and muscle, and almost destroyed some of our greatest food comforts in troubled times. No wonder we have waged such battles against them. They remind us of our animal selves and our mortality. These pestiferous insects are significant because of the ways in which they have shaped many of our attitudes and insect-related narratives more generally — narratives infused with fears and anxieties that now impede reasonable uses of insects for food and medicine.

Vincent M. Holt, in his 1885 pamphlet "Why Not Eat Insects?" replies "Why not, indeed! What are the objections that can be brought forward to insects as food?" He imagines the reply of his Victorian readers to be "Ugh! I would not touch the loathsome things, much less eat one!" Yet, he argues, eating insects is scientifically rational and socially reasonable, practiced all over the world for good reason. "What a pleasant change," he writes, "from the labourer's unvarying meal of bread, lard, and bacon, or bread and lard without bacon, or bread without lard or bacon, would be a good dish of fried cockchafers or grasshoppers." For some reason, Victorians were not immediately attracted to his menus, which included woodlouse sauce, curried cockchafers, and moths on toast.

Sixty-six years later, the entomologist F.S. Bodenheimer's comprehensive scholarly treatise titled *Insects as Human Food: A Chapter of the Ecology of Man* fared no better. Despite his having enlisted the support of Aristotle, Pliny, and Immanuel Kant to his cause, Bodenheimer's scientific and scholarly colleagues and students pretty much ignored his advice to eat bugs. The picture that opens the book — of a naked Australian Aboriginal woman carrying a baby, allegedly out looking for insects — may have titillated postwar *National Geographic* readers, but Bodenheimer's assertions that "primitive peoples" and "natives" had no qualms about eating insects did not convince a postwar European population hungry for meat, milk, eggs, potatoes, and gefilte fish. The book's colonial, condescending, patriarchal language is an unfortunate barrier to twenty-first-century readers; Bodenheimer was a professional entomologist with broad historical and practical interests, and, once one gets past the language, his book is replete with interesting anecdotal and scholarly information.

As Holt, Bodenheimer, and many since have discovered, one of the consistent responses of non-insect-eaters to insects on the menu is *disgust*, a word that has its etymological roots in Old French and Latin, *dis* meaning "the opposite of " and *gust* meaning "taste" (as in gusto and gustatory). The related term *revulsion* is rooted in the Latin for "pulling or tearing away," which would be the action we most often take when we are disgusted with something. Disgust and revulsion are visceral responses with evolutionary roots; they prevent us from eating rotten or disease-infested foods. Still, the fact that certain foul-smelling, rotten, worm-infested cheeses or rotten fish are considered delicacies shows that such foods can be an acquired taste; we learn that they will not make us sick. We trust those who prepare them.

In the emerging, eclectic, global culture of the twenty-first century, science and imagination converge and reinforce each other in unexpected ways. A tasty garnish of black ants on fresh salmon may unexpectedly evoke the 1977 film *Empire of the Ants*, or Stephen King's *The Mist*. Large-sized insects, of the griffenfly kind that went extinct a few hundred million years ago, are especially terrifying. Thus the bugs on my plate at le Festin Nu evoke the human-sized cockroaches in Burroughs's *The Naked Lunch*, David Cronenberg's *The Fly* (as well as the 1958 version), and *The Deadly Mantis*, a 1957 movie in which a giant, prehistoric praying mantis is freed from the polar ice and attacks humanity. In the twenty-first century, *The Deadly Mantis* might be viewed either as parody or as a morality play about the freeing of methane from Arctic bogs because of global warming. Seen as a product of its time, however, the film does not encourage philosophical reflection or public enthusiasm for reducing greenhouse gas emissions.

If you browse the internet for bugs — not unlike a goat browsing in an abandoned farmyard — you will come across websites and headlines that are grounded in personal or cultural disgust and designed to attract attention. Some of them are news stories of tiny bedbugs crawling into people's most private sanctuaries and sharing bodily fluids. Others have titles such as "10 Horrifying Insects That Will Make You Reconsider Ever Visiting Japan" and "Real Monstrosities." Still others mix the language of information with that of elevated anxiety, such as "Insect Swarms Plague Many Canadians Right Now. Here's Why," which was recently flagged on a weather site.

Sometimes the stories are not, in the first instance, told to create horror, yet the language used tends to promote fear. In 2014, the *Guardian* newspaper posted a link to close-ups of insect heads by Indonesian wildlife photographer Yudy Sauw. The images are colorful and strange, but rather than invite us to explore the unsettling and curious world of these tiny animals, the caption reads, "Face your fears: extreme creepy-crawly close-ups." Why was it framed this way, rather than as a photographic exploration of the micro-world, a curiosity?

Anne Raver's *New York Times* review of Amy Stewart's *Wicked Bugs: The Louse That Conquered Napoleon's Army and Other Diabolical Insects* appeared in the Home & Garden section of the paper. This placement speaks to how subtly cultural images inadvertently reinforce each other. Raver is interested in the invasive Asian stink bugs that are eating her tomatoes even as Stewart insists that she is only interested in bugs that altered the course of human history.

Stewart's catalogue of "wickedness" includes stories of the death and devastation created by black flies. They are reported to have killed twenty-two thousand cattle along the banks of the Danube in the 1920s. In the tropics, they carry a larval form of *Onchocerca volvulus*, a parasite that causes river blindness, which affects tens of millions of people annually; one of these flies, in Africa, bears the no-nonsense name *Simulium damnosum*. She also recounts the tale of the Formosan subterranean termites and their relationship to the failing of the levees in New Orleans during Hurricane Katrina. According to Stewart, "Gregg Henderson, the termite guy, raised the alarm about the fact that they were nibbling away at the seams of the flood walls [in the years before Hurricane Katrina]. But entomologists have a hard time in convincing other people that little creepy creatures can be so powerful."

F.S. Bodenheimer, in his comprehensive 1951 review, asserts that eating lice is "almost cosmopolitan." Human body lice evolved from human head lice a mere 100,000 years ago, although pubic lice (*papillons d'amour*, if you are French) appear to have been acquired — how, one wonders — from gorillas. In our post–gorilla intimacy urban societies, with our public health programs and emphasis on the links between godliness and cleanliness, we are less attracted to louse-nibbling behaviors, however cosmopolitan they may once have been. We are more comfortable with Stewart's brief, titillating history of the typhus-bearing body lice that brought Napoleon's army to its proverbial knees in Russia than we are with a history of humans actually eating the things.

Despite her disclaimer about focusing on bugs that changed history, Stewart devotes several pages to the historically marginal praying mantids and golden orb-spiders, whose females sometimes eat the males after (or during) mating — although she is quick to point out that "no bug is truly wicked. It is just eating." As May Berenbaum asserts, even "people who can't keep straight in their minds the concept that spiders aren't insects seem comfortably fluent with the notion that praying mantids are unreconstructed sexual cannibals." Berenbaum notes that sexual cannibalism has been reported, at least occasionally, in a wide variety of insect species ranging from crickets and grasshoppers to antlions and ground beetles. Despite their pride of place, only a handful of the 180 species of mantids have been reported to engage in the practice, and even then only sometimes, in some situations, and often under artificial laboratory conditions. The original description that launched the reputation of the predatory mantid was a 500-word story, in 1886, based on a single male and a single female kept as pets in a jar by a friend of the author. Berenbaum's take on the public fascination with this practice is that people "hate to let go of things sick and twisted," which can also be said for stories of human cannibalism. It is of course usually those in power, who feel their power threatened, or the invading, colonizing armies, who have accused their enemies (Caribes, Native Americans, Jews, Scots, Picts, most Africans, the Chinese) of cannibalism. When cannibalism was reported in Hannover, Germany, or Milwaukee, Wisconsin, or when a plane full of rugby players resorted to the practice after they had crashed in the Andes, no one suggested that all rugby players, or Germans, or Americans, were cannibals. Any inference made from these accounts that *all* humans might be secretly eating each other (*Soylent Green* notwithstanding) would be treated as a twisted, Monty Pythonesque joke.

This is not to say that Stewart is unusual in her mixed messaging. Even when the content of the message is, "they're not so bad," or even "they are useful," when it comes to insects, the marketing headlines often undermine the substance. In his book *Living Things We Love to Hate*, which in its content is trying to rehabilitate the public image of these animals, Des Kennedy nevertheless uses chapter titles such as "Flies — Awful Fecundity" and "Wasps — The Social Terrorists."

The pestiferous, disease-bearing bugs, which are but a tiny fraction of all those millions and millions out there, have given all insects a bad name. They instill fear and cause us to whimper. Can we bear to eat them? Given the intimacy between food and bodies, is this not a kind of satanic communion? That bad name now hinders our ability to see them as they are, in themselves. This is not unusual for people; we all, scientists as much as religious fanatics, see the world according to our preconceptions. As I write this book, rumors of one Syrian terrorist suddenly cast suspicion on all Syrians. One Christian ideologue, one Muslim terrorist, one atheist scientific bully, and suddenly all members of that group are branded with the same sizzling iron. If we can learn one good thing from considering insects as food, it should be the ability to pay attention, to see the category-transcendent and subtle beauties and terrors of the world as it exists; in the words of George Harrison, life is flowing on within you and without you.

It will not be enough, however, to change the images in our heads. Our views of pestiferous insects have influenced how we respond to them in very practical ways. These cultural images, reinforced by certain scientific and economic narratives, have shaped how we practice agriculture and fight diseases. And it is our agricultural and public health practices that now stand in the way of a global shift to entomophagy. Just what are those practices? And — if the (m)admen can engineer a mind-shift in our attitudes toward insects — are there alternatives?

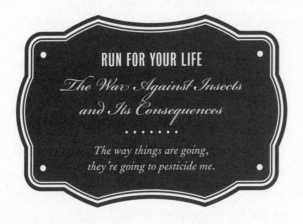

RUN FOR YOUR LIFE

*The War Against Insects
and Its Consequences*

• • • • • • •

*The way things are going,
they're going to pesticide me.*

"Millions of people owe their lives to Fred Soper. Why isn't he a hero?" So begins Malcolm Gladwell's 2001 *New Yorker* essay titled "The Mosquito Killer." In this essay, Gladwell recounts Soper's Global Malaria Eradication Programme, a post–World War II war on malaria based on what were seen to be the miraculous insecticidal effects of DDT. Soper was convinced that by spraying DDT in people's homes, malaria could be eradicated. Gladwell repeats the assertion that, according to some estimates,[58] DDT saved more lives between 1945 and 1965 "than any other man-made drug or chemical before or since." Paul Mueller, the Swiss chemist who discovered DDT's insecticidal effects in the late 1930s, received the Nobel Prize in 1948. By 1967, when Rachel Carson carefully documented the unintended consequences of profligate pesticide use, resistant strains of the mosquito were already emerging, selected for by the ways in which DDT was used.

The battles over pesticides run like a fault line through the entomophagy debates. These arguments often pit public health advocates working in tropical countries against environmental scientists in more temperate climates. It is the kind of thought-wrenching quandary one envisions being carried on in the lawless cage matches that pit climate change, environmental conservation, economic development, agriculture, and food security against each other.

In their 2013 review of the new entomophagy movement, "How Then Shall We Eat? Insect-Eating Attitudes and Sustainable Foodways," Heather Looy and her colleagues recount the story of Sanambele, a village in Mali. Here, children once foraged for grasshoppers as part of a diet that included millet, sorghum, maize, peanuts, and fish. When Malian farmers switched to growing water-hungry and pesticide-dependent cotton, they made more money, but at the cost of increased protein-energy malnutrition. Similarly, in Madagascar in 2012, the locust plague pitted those who wanted to eat locusts against those who wanted to defeat and kill them with pesticides because they were destroying familiar, staple food sources. It is a conundrum that ripples through all the fissures in the debates about eating insects and could, if not handled well, bring down the whole entomophagical edifice.

Because arguments about pesticides strongly intersect with entomophagy, it is worth examining the issues more closely. Since the advent of the Industrial Revolution and what President Eisenhower of the United States called the military–industrial complex, the weapons used to kill insects, like those used to wage war against other people, have increased in their lethality and the extent of their indiscriminate collateral damage. Just as aerial bombardment rains destruction on soldiers with missile launchers and shopkeepers selling zucchini alike, so DDT does not distinguish between edible crickets and deadly assassin bugs, between bees bearing honey and mosquitoes carrying *Plasmodium falciparum*. At the same time, the justifications for these wars have become treated as common-sense knowledge. Implicit in the twentieth-century use of pesticide sprays and flea bombs has been the assumption that of course we need to destroy them — or they will destroy us. Whatever we think of the religious superstitions of previous centuries, people in those times seem to have at least *considered* the possibility of other perspectives.

Within the non-bug-eating traditions of Europe, visitations of insect pests were often framed as moral problems. Locust swarms might be seen as attacks from Satan, or as armies sent by God to punish his wayward people. The latter seems to have been the Islamic view, but the Christian view could go either way. If a locust plague was God's punishment, then one would be called upon to suffer, repent, and change one's ways. This view aligned with the Greek tradition, which held that improperly expiated murder would lead the furies to send pestilence. In medieval Europe, the Church merely substituted demons for furies. This is the question that Father Paneloux struggles with in Camus's novel *The Plague*. If disease and pestilence are God's punishment for bad behavior, then is medical treatment a fight against God?[59] If insect plagues were the devil's handiwork, then this would create a more complicated, Manichean good–bad theology — why would an all-powerful God allow this suffering?

— but an easier practical solution. One could fight back with whatever weapons one had at hand.

In 880 CE, Pope Stephen VI provided holy water to exorcise a swarm of locusts in the area around Rome. More often than a direct papal decree, however, the decision as to the proper response to killer bee and locust infestations was decided by ecclesiastical courts with lawyers appointed to represent each side. Elaborate legal cases were initiated, with lawyerly machinations and florid speeches that could have been lifted from the O.J. Simpson trial.[60] These legal wranglings were not much different from class action suits and political debates over pesticide use in the current century. It's just that now, when we argue about whether such products foster or destroy sustainable food security, we talk about whether they are working with or against "nature" — a modern, ambiguous stand-in for the notion of a god.

The advent of industrially produced pesticides did not change this framing of the war against insects, which remained embedded in concepts of good (what is "natural," or, alternatively, what enables us to produce more food) and evil (wild nature, forces that destroy humans and human food, anti-technology Luddites). But the way these forces have been interpreted, especially in the debates around overpopulation and overconsumption of earth's resources, is sometimes flipped. In the 1990s, I was a member of an online discussion group about sustainability. We were all scholars or professional scientists of one sort or another. I was somewhat taken aback, then, when one member of our group, on learning that I was an epidemiologist, announced that "my people" were the problem. Epidemiologists had found and implemented too many ways to keep too many people alive! I was used to environmental issues being characterized as battles between good and evil, but I was disconcerted to discover that I was seen by some as being on the side of evil.

What changed with industrialization, and with the invention of DDT in particular, was the power and reach of our weaponry. During World War II, DDT dusting, bombing, and spraying averted catastrophic typhus epidemics of the sort that brought Napoleon's army to its knees, and enabled beachheads to be gained and battles to be won. The initiatives to control or eradicate malaria are often described in terms of discipline, battle, and war.

According to Gladwell, Fred Soper "was a fascist — a disease fascist — because he believed a malaria warrior had to be." He concludes his *New Yorker* essay this way: "There is something to admire in that attitude; it is hard to look at the devastation wrought by H.I.V. and malaria and countless other diseases in the Third World and not conclude that what we need, more than anything, is someone who will marshal the troops, send them house to house, monitor their every movement, direct their every success, and, should a day of indifference leave their shirts unsullied, send them packing."

On the one hand, using war metaphors in relation to how we treat disease offers some useful analogies for emergency medical care — quick and effective interventions based on knowledge, expertise, and skills. On the other hand, the use of such metaphors lumps together two very different kinds of activities and creates confusion that, in the complex web of eco-social life, can have profoundly detrimental outcomes.[61] This is especially so if the metaphors are confused with reality. The association of insecticides with modern warfare is even more tightly bound than that of medicine and war. Indeed, the marriage of war and pesticides has been sanctified in both metaphor and practice. PDB (paradichlorobenzene), a by-product of the manufacturing of explosives in World War I, was later used in pesticides and mothballs. Without World War II, it is doubtful that DDT would have so quickly gone from the lab to the field. Various formulations of arsenicals, organophosphates, hydrogen cyanide, and chloropicrin have likewise been used as weapons of war against people and insects.

Some would argue that we really do have wars against disease, with a medical armamentarium, just as we really do battle against insect pests. Airplanes and the internet are also technologies of military origin; and do we not value these products? War is where the money is. Why not take the products of war and put them to good peaceful uses, beating swords into ploughshares, bombs into power plants? I will not engage here in the more general debate on where technological inventions come from and how they are used. My concern is with how we frame our relationships to insects in general and the impacts this framing has on our desire to eat some of them. I'll not even try to make a case for saving parasite-bearing mosquitoes as a food source for bats, fish, and birds. Those cases represent unresolved, and probably unresolvable, questions among ethicists, ecologists, entomologists, and public health activists.

Insects, as a class and even as individuals, are a mixed bag, and, like people, the impacts of their behaviors can be seen as both good (they are sources of food for other species on which our food system depends, for instance) and bad (they are consumers of foods we would like to eat, and disease carriers). In the complex world we inhabit, it is often difficult to disentangle the two. Unfortunately, the battle against insect pests is predicated on the myth that we can do just that.

Heptachlor is an environmentally persistent organochlorine insecticide.[62] First registered for use in the United States in 1952, it was used against termites and several other insects that agriculturalists deemed to be pests. In 1976, in the wake of Rachel Carson's work as well as laboratory testing showing that heptachlor was toxic to the liver and reproductive tract in humans, as well as being carcinogenic, the US Environmental Protection Agency (EPA) banned all uses of heptachlor — except for use on seed corn to protect the grains during storage and for control of ants on pineapple plants. Ants, it seems, were protecting mealybugs because the mealybugs, like some other scale insects, including those that Yahweh offered the Israelites in the desert as manna, produced a sweet exudate that the ants valued. The mealybugs fed on pineapple plants. (Interestingly, since pineapples are *Bromeliaceae*, some of which are carnivorous and eat insects, this could be seen — not by Dole and Del Monte to be sure, but by some, perhaps — as poetic justice.) Heptachlor residues were not detected in the fruit of the pineapple plant, so this arrangement functioned without major hiccups for half a dozen years. Then, in 1982, health department chemists in Hawaii began detecting heptachlor in milk samples. Were these results correct? How could heptachlor get into milk? There followed the usual rounds of denials and retests and ambiguous statements suggesting that even if the pesticide was there, it did not appear to constitute an "unreasonable hazard." What happened was that pineapple growers, in trying to promote ecological efficiency, had devised a system whereby the pineapple plants were fed to cows as part of a nutritious diet. The growers were supposed to wait a certain amount of time between spraying the plants and feeding them to the cows, but neither cows nor farmers always follow written schedules.

Aldicarb is a pesticide that is effective against thrips, aphids, leaf miners, fleahoppers, and spider mites (which are not insects, but arachnids). On potato crops, it was used to kill soil nematodes. The active chemical in aldicarb inhibits the breakdown of cholinesterase, which is the chemical that normally deactivates acetylcholine at neuromuscular junctions after a message has been delivered from the nerve ending to the muscle. If acetylcholine is not broken down, you get convulsions. And sometimes respiratory failure. And death. In 1985, in California and Oregon, an outbreak of neurological disease (nausea, vomiting, abdominal pain, diarrhea, blurred vision, muscle twitching, slurred speech) was associated with consumption of striped watermelons. The symptoms were traced to aldicarb in the watermelons. The aldicarb had been (legally) sprayed on non-watermelon crops grown in nearby fields, or in the same fields the previous year. In 2010, the EPA announced an agreement with Bayer (the primary aldicarb producer) to phase out its production and use, stating that "aldicarb no longer meets our rigorous food safety standards and may pose unacceptable dietary risks, especially to infants and young children." According to the US EPA website, "Bayer has agreed to first end aldicarb use on citrus and potatoes, and will adopt risk mitigation measures for other uses to protect groundwater resources. The company will voluntarily phase out production of aldicarb by December 31, 2014. All remaining aldicarb uses will end no later than August 2018."[63]

At the same time, in 2011 AgLogic Chemical received EPA approval for Meymik 15G, another aldicarb-based pesticide, for use on cotton, dry beans, peanuts, soybeans, sugar beets, and sweet potatoes. The company's website says it is planning to "reintroduce aldicarb to the market for the 2016 growing season."[64]

A 2015 study of 225 wild and domestic animal deaths in the Canary Islands reported that 117 died of deliberate poisoning, more than three-quarters of these cases involving two pesticides — aldicarb and carbofuran, also a carbamate — banned in the European Union and Canada and restricted in the United States. The wheels on the bus, as my grandchildren used to sing, go round and round.

After all the bad public relations surrounding persistent organochlorines, many pesticide companies and producers shifted to fast-acting but less environmentally stable organophosphate pesticides. What this means is that there are fewer residues in the food, and hence fewer complaints from urban consumers, but much higher risks to poor farm workers.

Insecticides were created to kill insects: that's what the suffix -*cide* means, as in suicide, and patricide, and fungicide. Those who market pesticides would sometimes like to give the impression that they are being forced to shift among pesticides and to create new ones — all of which costs a lot of money — simply in response to the actions of irresponsible "environmentalists" who care more for a few insects or birds than they do for human health and food security. These environmentalists are never called *scientists*, which most of them are, even as their corporate challengers are paraded out as scientific experts — although where their expertise resides, exactly, is not always clear. The fact is, however, that reckless use of these powerful chemicals has selected for insecticide resistance in many species. This increased resistance to the drugs could have been predicted using basic evolutionary theory (apparently the pesticide producers only believe in science selectively). Whether the causes are greed or wishful thinking, we now have what we have. The question is, how do we respond?

For the companies who helped create the problem, the solution is easy: invent new and different, and more expensive, pesticides. As usual, each new class of chemical is paraded out as if it is a miracle. Neonicotinoids are a case in point. Somewhat cavalierly called neonics (nee-oh-nicks), even by their enemies, these chemicals are neuroactive in a manner similar to nicotine. Introduced in the 1990s to treat sucking, chewing, and soil-dwelling insects, they have rapidly become the most widely used pesticides worldwide, which would seem to contradict the notion that they represent a kind of miracle. Miracles, from what I understand, are unusual and rare. Neonics have been applied to most economically important agricultural crops, such as corn, canola, cotton, sorghum, sugar beets, rice, and soybeans, as well as to fruits and vegetables such as apples, cherries, peaches, oranges, berries, leafy greens, tomatoes, and potatoes.

Not many insecticides have been studied for their specific impacts on *edible* insects or insect products used as food. Recent studies of neonics on bees are an exception. These have been heavily researched because of their temporal and spatial association with the catastrophic decline of honey bee populations in many parts of the world, and more specifically with the emergence of colony collapse disorder (CCD). CCD is a syndrome in which all the adult bees simply abscond, leaving behind a queen, some honey, and a few immature bees. There are no dead bodies.

After more than a decade of intensive research, the general scientific consensus is that CCD is the result of interactions among land use changes, industrial honey bee management practices, multiple pesticides, certain species of varroa mites, bacteria (foul brood), fungal infections, viruses, and the immune systems and behavior of bees. Neonics loom large because of their ubiquity and, one might say, profligate application. As well, there is direct evidence of decreased sperm counts in drones exposed to neonics. Neonics are not alone among pesticides in being a problem; a survey of beehives in Canada and the United States, published in 2010, found evidence of 121 different pesticides and their metabolites in wax, pollen, and bees.

The decline and disappearance of our most celebrated and valued insect companions can be reasonably attributed to the interactions between the human war on insects and the industrialized systems we have designed to promote food production and security. According to Ernesto Guzman, a professor at the University of Guelph who has spent his career studying bees, "What's killing bees are modern practices of beekeeping and agriculture."[65]

As other non-bee insects move from being a local, indigenous food choice to being a commodity in the global trading system, can we evade the trappings and traps of industrial agriculture? Can eight or nine billion people live on this planet and eat in ways that will enable us to achieve food security in the larger context of all those elusive, ethically grounded global efforts we have set for ourselves: Sustainable Development, Sustainable Livelihoods, One Health, EcoHealth, Health-for-All, and Social-Ecological Resilience? Can we manage the conundrum we face when we want to impede the bugs that eat our food and inject us with parasites while at the same time promoting those we wish to eat? Is there a possible road to peace in this endless, self-defeating war on bugs?

## PART IV. BLACK FLY SINGING: REIMAGINING INSECTS

Of course, we know that insects are not *only* our enemies and destroyers. We have alternative stories. Who are the good insects in the stories we've told ourselves, and what can they teach us? Do they suggest options for more convivial policies? Can we do some kind of feminist narrative therapy on the human race that helps us find health and redemption? Let us explore the alternatives to our perpetual war on bugs.

MOTHER MARY COMES TO ME

*Insects as Creators
and Bodhisattvas*

. . . . . . .

*We're in love and it's a buggy day.*

Good nutrition. Ecological sustainability. Fewer greenhouse gas emissions. Are those black flies singing? Is this a fantasy?

In the United States, Black Friday, the day after American Thanksgiving, is a frenzied, greed-driven, chaotic shopping day. In 2012, Doug Currie, Vice-President, Department of Natural History and Senior Curator of Entomology at Toronto's Royal Ontario Museum, launched Black Fly Day as an antidote to the toxic commercialism of the day. Currie's 1988 Ph.D. dissertation was on black flies, and he has done extensive research into the diversity and biogeography of northern Holarctic black flies. His book on the Simuliidae (black flies) of North America, co-authored with Peter Adler and D. Monty Wood, won the 2004 Association of American Publishers Award for Best

Single-Volume Reference in the Sciences. So the proposal for a Black Fly Day was not necessarily a frivolous pun. But still, one might ask: why would one want to celebrate such a terrible pest?

Black flies are infamous in many parts of the world, mostly because of a few blood-sucking, death-dealing, river blindness–bearing black sheep (if black flies may indeed be said to have black sheep) in the family. But some of them can be seen from a more ambiguous perspective. Globally, humans have been rapidly blundering into and destroying many of the earth's most important and stunningly complex ecosystems. Who has protected the few remaining natural refuges from the depredations of humans? In the Arctic, in those few weeks that aren't butt-freezing cold, those great eco-warriors have been black flies.

In North America, home to slightly more than a tenth of the approximately 1,800 species of black flies, they are mostly nuisances with benefits. The larvae are fastidious and only live in fresh, flowing, oxygenated, pollution-free water; so if you see them when you go swimming, it's a *good* sign. The males drink nectar and pollinate flowers; they are the Ferdinands — the flower-smelling non-fighting bulls — of the bug world. Four species of male black flies have given up sex altogether, and the females reproduce parthenogenetically. In those species where the females have a taste for mammalian blood, they prefer nonhumans. Females from eight of the nine species restricted to the Canadian tundra don't even have the necessary blood-feeding mouthparts.

In 1979, our family drove from my first veterinary job in northern Alberta to my second job in what Albertans called the Banana Belt, 110 kilometers (about 70 miles) north of Toronto. When we stopped, after several hours of driving through endless boreal forests, for an idyllic view of Lake Superior, my two-and-a-half-year-old son came back to the car covered in blood. He hadn't even felt the bites. I guess there were a lot of hungry females out there unable to find nonhuman mammalian alternatives. On the plus side, the bites indicated that the lake water was fresh and clean. Some have even suggested that black flies were instrumental in "guiding" caribou along certain paths through the landscape, facilitating what we now call food security for indigenous people. Certainly, these insects are an important feed ingredient in the food chain, linking micro-algae to fish, birds, and thin-hided wandering primates.

The alternative narratives to war without end have existed, often quietly, alongside and often interwoven with the stories of bugs as pests and disease carriers. These alternative narratives provide opportunities to reimagine our relationships with bugs, and to invent or reinforce responses to six-legged afflictions, pests, and plagues that are more compatible with entomophagy.

We can begin by noting that for every online description of "real monstrosities" or entreaty to "face your fears," there are websites with more endearing titles, such as The Unexpected Beauty of Bugs and Beautiful Bugs of Belize. To counter *Alien*, we could mention *Wall-E*, in which the world-saving robot kept a cockroach as a pet. Others might recall the adventuresome grasshopper, centipede, earthworm, spider, ladybug, silkworm, and glow-worm who are James's traveling companions in Roald Dahl's *James and the Giant Peach*. Although honey bees are almost universally praised, ants are not without their cultural champions either. From the wisdom of Solomon in the *Book of Proverbs* to the movies *A Bug's Life* and *Antz*, Formicidae are raised up as exemplars of cooperation and hard work. Occasionally, as in the 2015 Marvel superhero film *Ant*

*Man*, they are even heroic. The hero of the children's book *The Cricket in Times Square* is an endearing little fellow, and although the talking cricket in *Pinocchio* is annoying, he's really not a bad guy, and he probably didn't deserve to have his head smashed in by a thrown mallet in Collodi's original tale.

How do we begin to make sense of these conflicting cultural images and the scientific and cultural entanglements from which they have emerged? How do we emotionally and intellectually create a mash-up of *Alien* and *Antz*, malaria and edible beetles, river blindness, black flies, and clean water? How can we begin to cope with our emotional ambivalence about digging into a bowl of live termites, or outright revulsion at watching Star Trek's Klingons dip into a writhing bowl of Gagh?

A first step would be to recognize the biases in our own narratives and not just the flaws in others. Despite the brilliant and Herculean efforts of that great Swedish naturalist and polymath Carl Linnaeus to standardize our descriptions of living things, even the most hard-core of hard scientists still fall back on culturally based metaphors and stories, if not to describe the things themselves, then at least to talk about their roles in nature. These metaphors and stories influence how we think about living things and, in turn, whether we wish to eat them or not. Do assassin bugs, as the name implies, kill important leaders for political or religious reasons? Or are they merely insects that kill and eat other insects? Is the large female insect — the one who carries the eggs and determines the genetic makeup of the beehive or ant colony — a queen? Surely not one that Elizabeth of England or even Lewis Carroll's Red Queen would recognize as such. Similarly, the use of the terms *workers* and *soldiers* for ants, termites, and bees reflects political and social histories in England and India.

Social insects such as ants, bees, and wasps are particularly susceptible to having had colonial imaginations imposed on them. During World War II, ethologist Karl von Frisch, whose maternal grandmother was Jewish, was allowed by the Nazis to continue his work on honey bees despite other researchers classified as *mischling*, or mixed-blood, having been forced out of their jobs. Ernst Bergdolt, editor of *Zeitschrift für die gesamte Naturwissenshaft* ("Journal for the Entire Natural Sciences"), tried to get Frisch removed from his post at the Institute of Zoology in Munich. Bergdolt did not believe that Frisch was sufficiently cognizant of the ways in which bee society, so systemic and well-organized, could be seen as a model for a Nazi utopia. For Frisch, the bees were simply his friends, a refuge from the violent chaos around him, evoking awe and a sense of reverence for nonhuman life. In 1973, Frisch received a Nobel Prize for his discovery of the complex communications and decision-making practices of bees. Does that make Frisch's loosely organic notion of the hive the correct one? Or, perhaps, as apiarist and Buddhist monk Michael Thiele has asserted, "honey bees are bodhisattvas," who "mirror our own struggles to live in the world" and, in their instinctual wisdom, inspire us to "new ways of living."

Bert Hölldobler and Edward O. Wilson, in their book *The Superorganism: The Beauty, Elegance and Strangeness of Insect Societies*, write that decision-making in a beehive "is a highly distributed process of friendly competition among the scout bees that identifies the best site. It is, in effect, a democracy."[66] Ah, so now we know. But is it a parliamentary democracy, with a queen? Or the republican form more familiar to those American authors?

Not content with imposing their metaphors on European science, colonial empire-builders exported their beliefs to many countries around the world, so that in adopting the "scientific method," African, Latin American, Vietnamese, and Japanese scientists have been contaminated by the linguistic mindsets of nineteenth-century colonial, Royal Society Europeans.

For those who are promoting entomophagy, the cultural baggage carried by insect names are more than curiosities for cultural critics and anthropologists. They create some quandaries. Bees are excellent protein sources if eaten directly — as good as or better than crickets and mealworms. Yet while Westerners may be quick to adopt crickets and mealworms, and harbor no moral qualms about eating hornets and wasps, they may balk at a curry in which baby honey bees are the main ingredient. Is it because we secretly believe in bee bodhisattvas, or extol the virtuous lessons of governance and democratic socialism the hive offers? Do not ants, termites, and hornets offer similar lessons? Is it because bees seem "cuddlier," more like pandas than grizzlies? Or is it because bees are now deemed a critical component in industrialized monocultures? It is, I am guessing, a complex, confusing mix of these things.

Some alternative narratives weave together strains from insect-eating and non-insect-eating cultures, drawing on the best of both, and suggesting new pathways.

The songs of insects have been one area where such cross-cultural conversations about the mixed blessing of insects have been rich and relevant to the study of edible insects. Hungarian composer and entomologist Béla Bartók mimicked cricket sounds in his 1926 suite *Out of Doors*. Bartók apparently felt that collecting insects, like collecting folk melodies, was a responsibility for contemporary composers. In 1979, American artist Jasper Johns created *Cicada*, a cross-hatched screenprint that evokes the complex, colorful, surround-sound song of the cicada. Inspired by Jasper Johns, South African composer Kevin Volans composed *Cicada*, a minimalist piece for two pianos. Poet Andrew Hudgens, in a poem, also titled simply *Cicada*, calls the insects an "oracle of our mortal summers" and a "song above our heads / in our hot corporeal evenings."

*Bug Music: How Insects Gave Us Rhythm and Noise* is a celebration of insect singing by philosopher and jazz musician David Rothenberg. In both poetry and hard-core musicology — involving the creation of music notation and technology suitable to recreate the sounds they make — Rothenberg explores the strange, beautiful, and mysterious music of cicadas, crickets, and throat-singing katydids. Meanwhile, on the website of Mr. Fung's Cricket Orchestra, Swedish cricket musicologist Lars Fredriksson — also known as Fung Liao, the composer and conductor of the cricket orchestra — introduces the "Chinese Cricket Rosary Ensemble," which "usually consists of 108 outstanding singing crickets of species like Bamboo Bells, Purple Bamboo Bells, Heavenly Bells, Golden Bells, Small and Large Yellow Bells." Fredriksson describes performances by his orchestra as having "a slight resemblance of the Wiennese heurigen, the Oktoberfest around München, and the testing of the Beaujoulais noveu [*sic*]."[67]

One American blog has gathered dozens of musical pieces and pop songs written about insects, including pieces about ladybugs, moths, butterflies, cockroaches, dung beetles, black flies, dragonflies, and crickets. They are not all, to be sure, celebrations, but they are not puke-inducing, scary songs. In popular culture there have been Buddy Holly and the Crickets, Iron Butterfly, and Alien Ant Farm — as well as, of course, the Beatles.

Taking things one step further into the territory of ento-mophagy, the Anderson Design Group in Nashville, Tennessee, has a web page devoted to cicadas that explicitly brings together some of these strains of food, fear, and insect music.[68] The web-site's title is the alarming Cicada Invasion, but the banner below is a more inviting "Sing. Fly. Mate. Die." The page itself includes rec-ipes (sixty-six, the last I counted), videos, stories, and pictures. The site poignantly declares that we "often forget that the life cycle of the cicada is both beautiful and tragic and just focus on the noise."

Attitudes toward insects in Europe have more often been rooted in religion, pornography, entertainment, and poetry rather than entomophagy. However, even here one might find possibili-ties for reframing the cultural imagination, linking them in our minds to love and entertainment.

Fleas aren't merely annoyances to pets and people, who also transmit the plague bacillus. John Donne's erotic metaphysical poem "The Flea" takes the form of a seductive plea to a female listener, describing the travels of a flea, first sucking blood from him and then wandering over to suck from her, thus mingling the pair's bodily fluids inside it. Blogger Bridget Lowe, under the title "Fleas Are for Lovers," suggests that readers "keep in mind here that the printed 'S' at the time Donne wrote his poetry looked more like an 'F' — allowing the poet to play a bit and still claim total innocence."[69] Donne's sensual poem was a sub-lime example of a less-than-exalted tradition of what some have called insect pornography, in which male poetasters commented more explicitly about the skitterings of insects into fleshy cleav-ages and under skirts.

Flea circuses, once reported to be extinct, have in fact been kept alive by Colombian-born artist María Fernanda Cardoso. Her *Cardoso Flea Circus*, which is now part of the permanent collection of the Tate Gallery in London, includes cat fleas trained to escape (Harry Fleadini), lift cotton balls (Samson and Delilah), walk tightropes (Teeny and Tiny), and pull a toy locomotive (Brutus).

The mixed reputations of insects cut a wide swath across the domestication spectrum. Silkworms are fully domesticated, completely dependent on human care. While they are brought together into teeming masses, they are not self-organizing the way bees are; we have used their cocoons for clothing and their larvae for food. Crickets are not domesticated, but have been prized as food and for their fighting, singing, and — if Walt Disney and George Selden (the author of *The Cricket in Times Square*) are to be believed — storytelling skills. Locusts are not domesticated at all; they are wild animals that have periodically visited plagues of biblical proportions upon evil enemies and offered spiritual blessings to hungry prophets in the desert. Bees are the semi-feral cats of the insect kingdom. From the Egyptians to the Mayans, from Minoan-Mycenaean goddess worship to Hinduism and Catholicism, *Apis* species have held an honored place in the mythologies of humanity. This tradition has continued well into the twentieth and twenty-first centuries. Novels such as Gail Anderson-Dargatz's *A Recipe for Bees* (1998), Sue Monk Kidd's *The Secret Life of Bees* (2002), and Laline Paull's *The Bees* (2014); movies such as *Bee Movie* (2007); and popular science and naturalist books such as Candace Savage's *Bees: Nature's Little Wonders* (2008) and Mark Winston's *Bee Time* (2014) are ample evidence of the special place bees hold in our cultural imagination.

It is one thing to raise up and magnify positive images of insects in science and culture. The great challenge for those wishing to invent a sustainable food supply that includes insects will be to find ways to acknowledge the bad along with the good, and to dance cleverly with the tensions that emerge. Indeed, unadulterated cute and good stories about insects might discourage entomophagy as much as those that characterize them as unrepentant and evil marauders.

"Run for Your Life" and "I've Just Seen a Face" are both acoustically accompanied songs from the Beatles' *Rubber Soul* album. "Run for Your Life," a possessive, jealous fit of male rage, was Lennon's least favorite song, one that he regretted having written. "I've Just Seen a Face," with its forward-propelling "I have never known / The like of this / I've been alone / And I have missed" is the acoustic Beatles at their best. Taken together, they encapsulate the ambivalent relationship between people and insects. *If we can't control you, we will kill you*, on the one hand, and *OMG you are so uncontrollably, wildly beautiful* on the other.

Martin Heidegger was a German philosopher whose 1927 book *Being and Time* is considered one of the most important philosophical works of the twentieth century. He was also, at least until the mid-1930s, a Nazi. Similarly, Ezra Pound, one of the most influential poets of the twentieth century, was a fascist. Karl Marx, that great champion of a future egalitarian society, lived a less-than-exemplary domestic life and is said to have abused his housekeeper. One might be tempted (as I have been) to separate the ideology from the philosophy and poetry. When I posed this dilemma to philosopher Karen Houle, she urged me to think about such people as complex and contradictory human beings, as we all are. This became, for me, a moment of awareness as to how we might imagine people and insects in more complex ways.

The idea that insects, bacteria, and people are either good or bad is one of our most dangerous illusions. What used to be called our logical left brain tells us insects are incredibly useful and mostly good for us. *Let's eat them!* Our intuitive right brain imagines that insects are monsters. *Let's kill them!* A bug observing this conflict might suggest that, as far as her fate is concerned, the right brain–left brain argument is moot. She will die in either case. Besides which, Roger Sperry's theorization of right brain–left brain split is now considered to be an over-simplification of a kernel of truth. People function best if the different parts of the brain work together; the corpus callosum, that bundle of nerve fibers connecting the two hemispheres of the brain, is what makes us complexly, fully human. As in ecology, the conversations among parts are as important as the stuff communicating. In promoting entomophagy, we forget these complex relationships at our peril.

Locusts are a deadly plague. They could also be part of a healthy diet. They are both, and their contradictory nature, like those of Heidegger, Pound, and Marx, is central to who they are. In "Penny Wiseguys," the 513th episode of the long-running television show *The Simpsons*, Lisa takes up eating locusts to combat an iron deficiency attributed to her vegetarian diet. Later, taunted by bugs in her dreams, she changes her mind, and, after a few misadventures she releases them — whereupon they immediately raze a corn maze.

From the possibly mythical folk storyteller Aesop (sixth century BCE) to modern variations of the ant and the grass-hopper tale in movies such as *A Bug's Life*, grasshoppers have been characterized as lazy and ants as hard workers. In the original tale, and for several centuries after "The Ant and the Grasshopper" entered popular literature, the grasshopper was actually a cicada, which, given its propensity to sing, makes sense. Over the centuries, perspectives on the ant and the grasshopper flip-flopped. In Somerset Maugham's short story "The Ant and the Grasshopper" (1924), the layabout brother marries a rich widow. John Ciardi's adaptation of the tale, *John J. Plenty and Fiddler Dan* (1963) — in which Fiddler Dan marries an unconventional ant — celebrates poetry over fanatical work, and John Updike's wastrel Brother Grasshopper leaves his hard-working but lonely brother a rich trove of memories.

Chinua Achebe, in his great novel *Things Fall Apart*, describes a scene in which "quite suddenly a shadow fell on the world, and the sun seemed hidden behind a thick cloud. Okonkwo looked up from his work and wondered if it was going to rain. . . . But almost immediately a shout of joy broke out in all directions. . . . 'Locusts are descending,' was joyfully chanted everywhere. . . . For although locusts had not visited Umuofia for many years, everybody knew by instinct that they were very good to eat." Later in the novel, the locusts become a symbol of the destructive swarms of white men coming into the country.

In Japan, often heralded as a leader in entomophagy, insects are pests, poets, and pets. From the song "Hotaru no Hikari," about a fourth-century Chinese scholar studying by the light of fireflies, to the anime "bugmaster" Mushishi, from Edogawa Ranpo, a writer who used insects to invoke horror in his stories, to Kawasaki Mitsuya, who aspires to heal familial relationships through having parents and children connect in new ways by caring for, and thinking about, stag beetles,[70] insects may be portrayed as both good and bad, but they are certainly woven into the cultural fabric.

In John Vernon Lord and Janet Burroway's children's story *The Giant Jam Sandwich*, the town of Itching Down (which is not "a waspish sort of town") is plagued by a swarm of four million wasps. The townspeople try all the usual spray-and-swat killer responses, none of which work. Finally, Bap the Baker rallies the townspeople into a great community project: to create a giant jam sandwich in which to trap the wasps. In the end, the townspeople prevail. At the end of the story, the giant jam-and-wasp sandwich provides a feast for birds "for a hundred weeks."

An updated version might have the townspeople feasting on the sandwich themselves, but feeding the birds seems less selfish and more ecologically appropriate. In any case, the key to Itching Down's solution to the wasp problem is that it targets only the ones that are pestering the town and uses natural animal behaviors to get rid of the pests. No nerve-gas weapons of war involved.

*The Giant Jam Sandwich* got me thinking about how we might celebrate Black Fly Day. I see it as a general celebration of insects, a day to give thanks for the pure water that black flies alert us to, for the crickety biscuits in the oven, and for the awesome and awful locusts in the wild. What if we had a grand celebratory feast and invited Aboriginal Australians, indigenous people from Africa and Amazonia, China and Southeast Asia, as well as farmers from Ontario, Saskatchewan, and Nebraska? What if we asked each of them to prepare a meal that included insects or insect products, or products that depended on insects, for instance, for pollination? What if we explored the dark side of bugs, the crop pests and the malaria mosquitoes, even as we ate crickets or mopane worms or palm weevil larvae, pollinated nuts, cereals and fruits, bread with honey? I suspect we would not all agree, and that not everyone would be comfortable eating bugs; but that, in my view, is not the point. Perhaps we can begin to modify the cultural narratives and foods we use to define ourselves. The point is to begin to understand ourselves, and the world we inhabit, in its rich diversity, just a little better.

CAN'T BUY ME BUGS
*A New Age of Negotiation*
• • • • • • •
*The ladybug beetle is a pretty fine bug*
*(and she really has a lot to say)*

No amount of money can compensate for millions of dead insect species. Money can't buy me love, pollination, or complex, dynamic relationships among insects, plants, soil, and greenhouse gases. When insect species disappear, the magical mystery *Magicicada* musical will be silenced, and the trees, turtles, fish, and birds will suffer as they lose that periodic extravagance of fertilizer and feed. The insectivorous birds will disappear. Flowers will bloom once and then wrinkle and waste away. Once honey bees — or monarch butterflies, or dung beetles — are gone, shareholder profits will not bring them back. In the pesticide and fertilizer whorehouses, money can buy a one-night stand, a few seasons of corn or soy or canola. Pesticides provide temporary, short-term, transitional satisfaction for managing our culinary desires.

In a 2011 scholarly review titled "Energy-Efficient Food Production to Reduce Global Warming and Ecodegradation: The Use of Edible Insects," environmental engineer M. Premalatha wrote, "The supreme irony is that all over the world monies worth billions of rupees are spent every year to save crops that contain no more than 14% of plant protein by killing another food source (insects) that may contain up to 75% of high quality animal protein." The global agri-food system, however — like the economy in general — does not run on irony.

How can we begin to reconcile, not just in our heads and hearts but in practice, our conflicting experience with, and mixed feelings about, bugs?

At first glance, it may seem that one strategy to control insect pests without using insecticides would be to eat them; after all, people already eat locusts. It is a crude strategy, and it has been tried. With a few exceptions, eating insect pests has not been very successful in controlling them. Still, in a search for nontoxic strategies to manage human–insect–food relationships, it is worth looking at those exceptions.

In Thailand, in the 1970s, the Bombay locust *(Patanga succincta)* — normally a forest-dweller — was becoming a serious pest in maize fields planted in forest clearings.[71] When aerial spraying of insecticide failed, the government promoted eating the locusts and even promulgated recipes. Today, deep-fried *patanga* is a popular snack, and the species isn't considered a serious pest. There are even farmers now who grow maize to feed the locusts, which bring a better price. So, are they food? Are they pests? Yes. And yes. And most assuredly, fifty years of exposure to eating locusts is better for one's health than fifty years of exposure to pesticides, no matter how low the residues.

About eighty species of grasshoppers and locusts are eaten worldwide. Although there is a lot of variation in their nutrient content, most locusts, with about 60 percent protein and 13 percent fat (dry weight), are right up there with cows and cockroaches as excellent sources of human nutrition. They are not, for many people, a "novel" food source. There is a long history of humans around the world eating locusts and grasshoppers. Studies of human feces at Lakeside Cave in Utah indicate that, at various times going back 4,500 years, hunter-gatherers near the Great Salt Lake sometimes ate locusts and grasshoppers. Millions of grasshoppers and/or locusts periodically crash-landed into the waters of the Great Salt Lake. Washed up on the shore, naturally salted and sun-dried, they became a grand buffet. More recent ethnographic and ethnohistorical studies reveal that grasshoppers and crickets were part of the diets of some indigenous people in the area well into the late nineteenth and early twentieth centuries.

So when I read of the locust plagues that devastated large parts of Madagascar in 2012 and for several years following, my mind drifted to the possibilities of fall-season soups and mock-raisin breads. I wondered whether people could have just eaten the locusts. Why not? Well, they could have, but it wasn't as simple as that. The voracious locust swarms destroyed rice fields and pastures, causing hunger and threatening the food security of thirteen million people. Making a bad situation even worse, the plague hit the news just before Passover, and hence, in the Judeo-Christian imagination of Western societies, resonated with biblical implications. It was both a devastating plague and a public relations nightmare.

International and government agencies sprayed insecticides

to manage the hungry pests, thereby contaminating a possible alternative food supply. Some children, though, were catching them by hand or in mosquito nets, drowning them, and roasting or frying them. Other farmers explained that the locusts might be a good source of food, but they did not keep as well in storage as rice. They rotted. There was no generic, one-size-fits-all response to the locusts. To adequately address the problem would have required facing the challenge of stopping the plague and, at the same time, developing new ways to harvest, store, and preserve the locusts for food. Given the cultural dynamics of postcolonial societies, and the sense of embarrassment that may accompany eating "bugs" in front of Europeans, such an approach would require a lot of courage and engagement with people where they lived, talking to farmers, elders, cooks, and children — and some major rethinking of strategies and appropriate, innovative technologies.

In Lockwood's description of the American locust plagues, he notes that some farmers were initially happy that their poultry were stuffing themselves on locusts. This happiness disappeared when their chickens and turkeys gorged themselves to death. The farmers tried to manage this lethal feasting by giving the birds a bit of grain before turning them loose on the locusts. But still, there were so many! Too many! More problematically, the farmers later reported that the flesh and eggs of these poultry were inedible, exuding a pungent, oily odor. Others lamented the great stench of rotting carcasses along the lakeshore and in ponds, streams, and wells. Again, one of the issues raised by this situation — an intense, special case of the issues faced by all human settlements for more than ten thousand years — was the question of how best to harvest and preserve sudden windfalls of food. This question has been a driving force behind the long histories of fermentation, salting, sugaring, refrigeration, drying, vacuum-packing, and, more recently, genetic modification of fresh produce to extend shelf life. I have the sense that, were we to take insects seriously as food, we could solve the storage and preservation problems as we have for grains, dairy products, and fresh produce. In pre-Columbian America, some indigenous groups came up with the ingenious idea of making a "desert fruitcake" of insects, pine nuts, and berries, mashed together and sun-dried. The Honey Lake Paiute prepared a soup of dried crickets and locusts. The Japanese have produced hornet pickles and alcoholic drinks, Europe has its history of mead, and a group in the United States is now testing beer fermented with yeasts carried by wasps. The possibilities may not be endless, but the list of preservation methods is most assuredly long.

In Mexico, some species of grasshopper are considered serious pests of corn, beans, alfalfa, squash, and broad beans. Since the 1980s, many farmers have tried to control them through spraying organophosphate insecticides (mostly parathion and malathion, both of which are considered *relatively* nontoxic to people). The grasshoppers are also recognized as a source of food, an Aztec tradition going back at least 500 years. Even today, between May and September, harvesters from Santa María Zacatepec (Puebla) head out into the fields before dawn; they are able to capture 50 to 70 kilograms of grasshoppers per week, and 75 to 100 tons per year. The annual sale of this grasshopper harvest brings in US$3,000 per family; for six months, this provides the main source of income for these people.

This is all well and good for the harvesters, but what about the farmers who want to control the pests? Two researchers from the Universidad Nacional Autónoma de México decided to find out. Over two years in the first decade of this century, René Cerritos and Zenón Cano-Santana monitored grasshopper infestations in field plots that had been sprayed and compared them to plots where grasshoppers were harvested manually. Although the lowest grasshopper infestation rates were in fields that were treated with insecticides, the researchers concluded that mechanical control still reduced the infestation to manageable levels, saved the farmers annual costs of US$150 for insecticides, brought extra income into the village, reduced risks associated with water and soil contamination, and eliminated negative effects on nontarget species.[72] Mechanical harvesting had the added social advantage that it required farmers and harvesters to talk to each other and coordinate their activities. The World Bank used to call this social capital, and, in a region where social breakdown is a problem, it is not a trivial advantage.

In the long run, we need these kinds of alternative commitments; complex eco-social systems are more equipped to resist pest infestations, but nurturing these systems will take some serious rethinking of how we live. In the meantime, can we find ways of living, however uneasily, together with insects? Although the Soviet Union and the United States never fought their ideologically based wars directly, the bloody battles in Guatemala, Nicaragua, Honduras, Uruguay, Angola, Mozambique, Cambodia, and Vietnam were surrogates. Non-Russians and non-Americans died in large numbers to keep alive the Russian and American dreams of world domination. That is how empires work. Similarly, since Rachel Carson's documentation of the unintended negative consequences of pesticides entered the public discourse, the war against pestiferous insects has not stopped, but merely shifted. These war metaphors, having informed medical practice, are now, in the language of surgical strikes, coming full circle, creating a mythology that offers the illusion of killing no innocent bystanders.

One of the most widely known and practiced strategies to control insect populations with minimal collateral damage is what has been called companion planting (by friendly gardeners) and intercropping (by more serious business farmers). More than 1,500 species of plants have some insecticidal properties, but even noninsecticidal plants can provide some field-wide resistance to the spread of pests. Another strategy is to bring in other insects that prey on or parasitize the ones you don't want (the pest-control equivalent of surrogate wars). More recent strategies have included the use of pheromones, genetic modification, and even playing distressing, infuriating music. I'll only talk about a few of these strategies to make the point that, even if some agribusiness leaders support the contestable and doubtful assertion that pesticides are necessary to feed the world, we have options other than starvation and revolution.

Farming of course predates industrial pesticides by more than a few millennia. A 2013 report from the National Academy of Sciences in the United States suggests that agriculture in China goes back more than twenty thousand years. Citrus trees have probably been cultivated for a couple of millennia. Mandarin oranges, which as children in the hinterlands of western Canada we called Japanese oranges, originally spread from their birthplace in North India or Southern China across Southeast Asia, and from there to Europe and around the world.

Having cultivated citrus trees for thousands of years in a country that was the birthplace of entomology, it is no surprise that Chinese farmers had experience with pests and nontoxic pest controls. They were aware, for instance, that the citrus stink bug, citrus leaf miner, leaf-feeding caterpillars, and aphids could — and would — attack their lemon, orange, pomelo, and tangerine trees. Not having access to malathion, acetamiprid, methidathion, cyhexatin+tetradifon, spinosad, and other modern weapons of the war on insects, they tried working *with* nature rather than running immediately into battle against it. Perhaps they had read the advice of Sun Tzu in *The Art of War*, written almost half a millennium BCE: "The supreme art of war is to subdue the enemy without fighting."

Chinese farmers are credited with instigating the first documented use of insects to control other insects. About 1,700 years ago, they discovered that a strain of weaver ants — yellow citrus ants (*Oecophylla smaragdina Fabr*) — would eat a wide variety of plant-eating pests. In the early years, they tracked down and collected nests from the wild; later (about 985 CE), using fat as bait, they trapped the ants in hog and sheep bladders. After about 1600 CE, the farmers discovered that, if they constructed bamboo bridges between the trees, the ants would occupy the whole orchard even if only a few of the trees were seeded. Winter was a challenge as the ants had trouble surviving the cold, so the farmers started collecting the ants in the fall and feeding them citrus fruits until the warm spring days returned. Finally, some observant farmers noticed that the thicker foliage of pomelo trees provided better protection — a sanctuary, if you will — for the ants. If the farmers had mixed groves of oranges and pomelos, and built bamboo bridges among them, the ants would nest in the pomelo trees and serve as an annual, renewable source of insect control.

In *Six-Legged Livestock*, a 2013 FAO report on edible insect farming, collecting, and marketing in Thailand, the authors noted that weaver ants were also used for pest control in mango orchards. Some farmers maintain their own nests, but finding queens and good sanctuary trees is a challenge, so often the ants are foraged. The farmers create ant highways between trees with rattan or cane ropes, which the ants — who are remarkable engineers — then use to move to new sites, where they build new nests from larval silk. Weaver ants are celebrated in songs and dances in northeastern Thailand, where their eggs, pupae, and adults are incorporated into salads and omelets. (Eating other pest-control products, such as insecticides, is not generally recommended.)

European and North American agriculture expanded most rapidly during the period when industrial pesticides were widely available and only minimally controversial. Non-insect-eating cultures have become addicted to these toxins and have aggressively marketed their drug habits abroad. Now, after decades of pesticide addiction, many agriculturalists in China, Europe, and worldwide are rediscovering "beneficial" insects.

Generally, the less toxic and more ecologically sustainable approaches to pest control — that is, those most compatible with entomophagy — require much more sophisticated agricultural practices and knowledge of ecology than using insecticides. In a 2016 report on the control of cochineal pests in prickly pear plantations in central Mexico, the researchers concluded that six different species of natural predators did keep the pest populations in check, as farmers had reported. They cautioned, however, that such "autonomous biological control" methods depended on agroecosystems with structural complexity and species diversity.[73]

Often, just to hedge their bets in the face of scaremongering by pesticide manufacturing companies, twenty-first-century farmers use a combination of natural predators and pesticides. Integrated Pest Management (IPM) considers and uses all forms of insect control, employing insecticides only at specific times in the growth cycle of crops. The medical–war analogy for this would be surgical strikes. Many of the IPM methods use natural enemies of insects, such as bacteria or protozoa. If seeded into standing water, for example, different strains of *Bacillus thuringiensus* will kill mosquitoes and black flies, and *Bacillus popilliae* will kill Japanese beetles.

Although René Antoine Ferchault de Réaumur recommended the release of lacewings into greenhouses to eat aphids as early as the eighteenth century, it is only in the past few decades that biological controls have gained some traction among non-insect-eating agriculturalists. Hundreds of species (and millions of individuals) of insects, including the infamous tiny parasitoid wasps, which are not interested in stinging people, are now raised by the millions around the world, specifically for release into greenhouses and onto field crops. Parasitic wasps can find the underground pupae of corn earworms and armyworms and lay their eggs on or in them; the wasp larvae then eat the worms. Commercial applications for the use of three species of wasp — *Diapetimorpha introita*, which attacks beet armyworms, *Cryptus albitarsus*, which attacks tobacco budworms, and *Ichneumon promissorius*, which attacks ten other insects considered pests in the United States — are currently under investigation. By 2000, there were more than sixty-five companies worldwide producing these "natural enemies," many of them for the greenhouse market.

Among North American farmers and gardeners, ladybird beetles were the test case that proved the value of biological controls. They also demonstrated the importance of fine distinctions. Ladybird beetles, or ladybugs as they are known in North America,[74] have been both celebrated and misunderstood. The celebration is evident in the name, which is an abbreviation of Our Lady's Bird. These beetles are said to have earned this name when medieval farmers, plagued by sap-sucking aphids on their crops, prayed to the Virgin Mary for help and were rewarded by visitations from these aphid-eating insects. Of some 250 names for these beetles, in 50 languages, 63 include some variation of *Virgin* and 52 some variation of *God*. May Berenbaum also notes less exalted names given to these beetles, such as "Cowlady" or "Bishop is burning," while Waldbauer cites their Hebrew name, which means "creature of Rabbi Moses."

The misunderstanding related to the importance of fine distinctions. Some ladybug beetle enthusiasts, thinking that Mother Mary had things in hand, weren't sufficiently mindful. In the late 1800s, the California citrus industry was being attacked by the cottony cushion scale (*Icerya purchasi*), a pest that had inadvertently been imported from Australia. Charles Valentine Riley, chief entomologist of the US Department of Agriculture (USDA), who had been an important player in the battles against locust plagues in the United States and in designing phylloxera control programs in European vineyards, had an idea as to what might work to control the scale insects. Circumventing travel restrictions for USDA employees, Riley had his assistant Albert Koebele designated as a US State Department representative at an international exposition in Melbourne, Australia. In 1888, Koebele sent back hundreds of live ladybug beetles (*Rodolia cardinalis*), as well as a parasitic fly, *Cryptochaetus icerya*, which were released into the orchards. The subsequent success of bringing cottony cushion scale under control was attributed to the imported beetles, and soon farmers from all over the country wanted some.

This success, however, coupled with the lack of entomological knowledge, has created misunderstandings and confusion among Just Plain Folks like me. A ladybug is not a ladybug is not a ladybug, so that importing generic virgins into your garden may or may not work, depending on whether they prefer beach weather or a bracing, yet temperate, cold, and how they imagine their ideal mates But mostly, it depends on their food preferences. There are some six thousand species of ladybug beetle worldwide, many with different eating preferences. Members of the subfamily Epilachninae, for instance, feed on plants such as squash. Gardeners, in trying to save their squash from the lovely ladies, have tried to control them using parasitoid wasps, which are now also being used to control scale infestations.

The downside of importing natural enemies into new territories is that the immigrant predators may develop tastes for other foods; as in any war, killing friends and innocent civilians — so-called "collateral damage" — is a major problem in the war on insects.

Breeding and releasing sterile insects is a technique that has shown greater promise when used by entomologists to control pests than when used by political leaders to control human populations. Using variations of this strategy, insects (usually males) are sterilized by radiation and released into the population of alleged insect pests. Like most nonpesticide methods, this strategy requires an understanding of the breeding behavior and ecology of the insects, as well as some pretty clear ecological boundaries. This technique works best on isolated populations, such as those on islands, insects that are fussy about the species they feed on, and insect species in which the females only breed once but the males are more promiscuous. One of the test cases for the sterile-male approach was the eradication of the screw-worm fly (*Cochliomyia hominivorax*) from parts of North America. The approach has also been used to control a few species of fruit fly. Japan successfully eliminated the melon flies *Bactrocera cucurbitae* from several of its islands between 1971 and 1993 by releasing tens of millions of sterile males.

In a technique that I suspect some humans might envy, the females of many species of beetle, wasp, and butterfly carry around bacteria that are transmitted to males during breeding. The bacteria kill the males; the females get to keep the babies and pass on the genes, but they don't need to worry about the males going off and sharing bodily fluids with other females. Entomologists have yet to determine the mechanisms for this selective killing, but, if discovered, they might be used as part of a pest-control program.

A variation of the sterile insect technique was reported by US researchers in 2012 and again in 2015. Using CRISPR/Cas9, a "cut-and-paste" technique for altering DNA, they developed a strain of mosquitoes completely resistant to infection with *Plasmodium falciparum* (the parasite that causes malaria) but otherwise, as far as anyone could tell, completely healthy and able to breed. These scientists expect that when they release their "brand" (their word, not mine) into the wild, the altered mosquitoes will completely interrupt malaria transmission in the areas where they are released.

I am not a big fan of lab-based genetic modifications, since they lack the contextual complexity, temporal reality checks, and scanning for unintended consequences that one sees with slower-moving breeding programs. What might be some of these unintended consequences? Historically, similar disease-control programs for blow flies and tsetse flies have had some success by releasing sterile males into insect populations. Over time, fewer babies are born, and the population of insects may be reduced or even disappear, at least within defined regions such as islands or valleys. These new proposals are different, however. The modified mosquitoes are still there, reproducing, but can't be infected. What if the parasite or virus is a natural limiting factor in the mosquito populations, as diseases in wildlife often are? Will removing the parasite increase the reproductive success — and population size — of the mosquitoes? With a larger population size, will they pick up other viruses or parasites to which they are still vulnerable and carry them? Having voiced these worries, if this work is successful and malaria is eradicated, I would have to be a pretty miserable curmudgeon not to celebrate.

In the meantime, researchers in Ethiopia, which is not an island, and where the high-tech, branded, solution would therefore be just another neo-colonial white elephant, have found that hanging a caged chicken near your bed significantly reduces the mosquito populations nearby. Apparently the mosquitoes don't like the smell. Not a magic bullet, perhaps, but disease prevention and dinner in the same package sounds pretty good to me.

Some newer suggested control methods are both imaginative and even more ethically problematic. Southern pine beetles and western pine beetles like similar trees, but the two species never inhabit the same individual tree. David Dunn, a musician and composer who has studied the acoustic ecology of insects, wondered what would happen if he played southern pine beetle sounds to western pine beetles. After all, in 1989, a ten-day blast of Alice Cooper, Van Halen, Styx, Kiss, Ratt, and Judas Priest drove Panamanian leader Manuel Noriega from his place of asylum at the Vatican embassy. So what would happen if you played annoying music to beetles? What happened was that the western males mated — and then tore the females to pieces. Dunn has explored this concept further, composing nonlinear, chaotic electronic music and then playing it to insect audiences. "Dunn plays such sounds back to the beetles," says David Rothenberg, "and they tear each other to shreds. What more convincing reaction to one's music could one hope for than that?" (Note to self: avoid concerts by David Dunn.) To my knowledge, insecticidal music has yet to be tried on a larger scale.

Dunn's bug-aggravating music is a reminder that all the ways in which insects relate to each other and with the world around them — sound, sight, scent, pheromones, magnetism — offer opportunities for humans to converse with them, to harangue them, to manage our interactions in such a way as to minimize the damage. Rather than weapons of war against insects, these are languages to encourage argument, conversation, and the live-and-let-live attitude that Jeff Lockwood calls entomapatheia.[75]

## PART V. GOT TO GET YOU INTO MY LIFE

If we in the non-insect-eating world wanted to get bugs accepted as a crunchy lunch, how would we go about doing that? Are they already here, and we just don't notice them, creeping onto our plates through the back doors, kitchen windows, and animal barns? Can we learn from the many non-European societies who have eaten insects for millennia? Let us go deeper into this new continent of entomophagy. Let us look more carefully at the ways in which insects can be food, feed, and perhaps even — can our new, kaleidoscopic eyes see this? — friends.

**LEAVING THE WEST BEHIND?**

*Entomophagy in Transition in Non-Western Cultures*

• • • • • • •

*Those bamboo worms really knock me out.*

The first time I knowingly and deliberately ate insects, in Kunming, China, it was a simple act of cultural deference. If you are a guest in another country, and they offer you bugs to eat, it is rude to decline the offer.

In January of 2010, Xu Jianchu, an ethnobotanist working with China's minority peoples, hosted a richly diverse feast for our working group on the ecological and social interactions that lead to the emergence of new diseases. The revolving table seemed to me like a Tibetan Wheel of Life — delicate, colorful, carefully arranged chaos. There, among the stir-fried, kebabed, and spiced-up flowers, roots, pods, leaves, lobster sushi, chicken, and pork, was a basket of crisp, deep-fried worms. *Omphisa fuscidentalis*, the bamboo borer, is a moth of the Crambidae family, but in the kitchen and on the plate, the larvae are simply called "bamboo worms." Struggling to overcome my Canadian squeamishness in the name of cross-cultural collaboration, I tried some. They tasted to me like french fries, but with small, crunchy heads. Xu, watching me with a smile, said he preferred them not quite as thoroughly fried, a little softer and juicier, and then regaled us with tales of all the foods he had eaten, which seemed to cover just about anything that had once wiggled or crept on the earth, or flown in the air, or swum in waters brackish or sweet. I neglected to ask Xu about the possible destructiveness of traditional foraging for bamboo and the viability of rearing worms in bamboo plantations.

There are a lot of other things I would still like to ask Xu Jianchu, perhaps over a cup of insect tea, made from the feces of vegetarian insects. A 2013 scholarly review paper suggested that traditional Chinese insect tea lowered blood lipids and had antihypertensive and hypoglycemic effects.[76] I am thinking now that I must go back to Yunnan and ask Jianchu about the ten species of cocoons eaten by the Wa people, and what the Tibetans do with *Cordyceps*, a caterpillar whose body is taken over by a fungus.

One could spend many lifetimes exploring all the other bugs eaten by minority and indigenous people around the world, who could teach all of us non-insect-eaters so much about diversity in the kitchen. If we think that normalizing insect-eating offers a way to reclaim our past and renew our identity as humans, would these insect-eating communities not be excellent sources for new ideas as to how to do this?

While our global agri-food technocultures build or patch walls and fuss over efficiencies of process and economies of scale, where else but from among these minorities will come the voices of renewal? If eating insects offers us a "last great hope to save the planet," in Daniella Martin's memorable words, where else can we find the sanctuaries, the lost or silenced voices capable of articulating the right questions?

Thomas Cahill, in *How the Irish Saved Civilization*, argues that after the rich and powerful Roman Empire collapsed under the combined weight of its own bullying sense of self-importance and invasions by Goths, Visigoths, and Vandals, it was from small, vibrant Irish monasteries that philosophy, Greek learning, ancient manuscripts, a sense of humor, and civilized ways of discourse reseeded the darkened European cultural landscape. Similarly, Cahill suggests that, even as our arrogant, globalized, profit-driven civilization collapses into religious and trade wars under the weight of its own inequities and iniquities, societal and human renewal is "germinating today not in a board-room in London or an office in Washington or a bank in Tokyo, but in some antic outpost or other"[77] — that is, from the margins.

I am willing, at least for the sake of argument, to accept Cahill's assertion, but I am left with many questions: for what are we seeking the margins? This is not the Roman Empire we are talking about here, with the Euphrates River and Hadrian's Wall to proclaim its borders. In 1973, social planners Horst and Webber proposed that there are some problems which, because of messy boundaries, complex interactions, and different perspectives on what constitute solutions, cannot be tackled using conventional problem-solving or scientific methods.[78] They labeled these wicked problems. Often, what we see as an individual problem is in fact a subset of a much bigger, messier, problematic situation, which technical problem-solvers are trying not to think about.

In a wicked situation, solving what appears to be a single problem as if it were independent of its context often generates outcomes that are more troublesome than the original dilemma. For instance, we might rid a country of malaria by paving over the swamps where the mosquitoes live, but the loss of water infiltration decreases the groundwater recharge and leads to long-term water shortages, and the radiant heat generated by the pavement adds to local heat islands and contributes significantly to regional warming. Ironically, pools of stagnant water on parking lots and in storm sewers — which are free of insect predators such as fish — are better habitats for many mosquitoes than healthy swamps. Or we might solve a food production deficit through economies of scale, thereby creating ideal conditions for epidemics and putting smallholder farmers out of work.

Thousands of books and papers on subjects ranging from software development and business management to GMOs, health care, and climate change have been devoted to strategies that could be used to "wicked" such problems. The most successful strategies appear to involve collaborative work, imagination, and recognition that there are no definitive solutions. To paraphrase the words of novelist Douglas Adams (in *The Long Dark Teatime of the Soul*), "We may not have gone where we intended to go, but we think we have ended up where we needed to be."

Entomophagy, at least insofar as it is seen as a solution to the problem of sustainable global food security, is one subset of a much larger, wicked, problem. Perhaps we can begin to tame this animal if we can name it and suggest some boundaries. So what might that big problem be? Implied but not always stated, since there are conflicting views on this within the entomophagy movement (which is one indication that this *is* a wicked problem), is that the current industrialized agri-food system — the same one that has given us such food bounty for the last century and made possible our much-celebrated and sometimes despised "way of life" — is that big, wicked problem.

Given that our "way of life" has become globalized, where can we find the margins of the problem? Should we look for a temporal boundary, sifting through historical and anthropological sources as foundations for a twenty-first-century entomophagy? Daniella Martin and others have argued that, if we want to eat a true Paleo hunter-gatherer diet, we should be eating bugs. I am not convinced that a modern food fad based on evolutionary survival will serve us well in the next millennium. Still, the notion of scouring wadis and gulches for overlooked oases of food in a world of eight or nine billion people is not entirely without merit.

The world today is a fundamentally different planet than when our ancestors first dug termites out of mounds or began arguing with bees over who should have access to their sweet treasure hoards. Yet there are remnants, tide-pools of living knowledge and practice that have so far escaped the tsunami of industrial progessivism. These eco-cultural remnants are relevant insofar as they have survived into the twenty-first century; they have enabled people to survive and even remain resilient in the face of radical environmental, political, cultural, and climatic changes. The boundaries we can identify here go beyond the temporal to a combination of culture and geography.

Globally, we can begin our searches in the small enclaves inhabited by marginalized people in Asia, Africa, and North and South America, enclaves where ecological and survival knowledge are sequestered, such as the American mountain valleys or

Mexican forest glens that Jeff Lockwood, in talking about how locusts survived between plague years, and how monarch butterflies survive today, has called sanctuaries. These are the places where the identification, husbandry, processing, and preparing of insects are now being preserved.

But before we do so, it is worth pondering why these practices have been marginalized in the first place, and thus what the challenges might be in mainstreaming them. The progressive, modernist notion is that they are simply inefficient, suitable perhaps for subsistence, but not for a modern, science-based world. There is little credible evidence of any sort that this is true. Many ideas, people, and cultures have been marginalized by an odd mixture of so-called enlightenment science (that is, the science of non-insect-eating cultures), religious enthusiasm, colonial arrogance, television, social media frenzies, pop-stars, and well-funded public relations campaigns. Social Darwinism was long ago discredited and discarded by (most) evolutionary biologists. Its ideas, however, continue to creep into many of our activities, including our practice of science and the programs and activities we continually create to "lift people out of poverty," or to promote health and sustainability. Just as the Victorian language we use to describe social insects is encoded with ideology, so too are the critiques from some skeptics of the new entomophagy movement. While insects may well have been eaten historically by billions of people, it is argued, they have done so because they were poor, and starving, and had no other options. Insects might have been good for subsistence, but can they improve global food security in any substantive way?

Tim Flannery, the Australian ecologist to whom I referred earlier, asserts that hunter-gatherers are quite capable of doing any of the jobs on offer in the modern world, but that the reverse is not true. Flannery goes on to cite research that demonstrates that our "tendency towards civilized imbecility has left its physical mark on us. It's a fact that every member of the mini-ecosystems we have created has lost much of its brain matter. For goats and pigs, it's around a third when compared to their wild ancestors. For horses, dogs and cats it may be a little less. But, most surprising of all, humans have also lost brain mass. One study estimates that men have lost around 10 percent, and women around 14 percent of their brain mass when compared to ice-age ancestors."[79]

So, we *can* surely learn from people whose ways of living have survived by being invisible to the global economy, and who may have some tips on how to compensate for our failing brain capacities that do not involve taking more drugs. But we need to be careful to do this learning in ways that neither marginalize them further, shaming them into driving their insectivorous practices "underground," nor suck them into the mainstream and disempower them even as we seek their inherited wisdom. It is a delicate conversation, burdened with centuries of colonial bullying.

The roots of these attitudes are deeply embedded in non-insect-eating societies. One small vignette from Adam Hochschild's heart-wrenching history *King Leopold's Ghost: A Story of Greed, Terror and Heroism in Colonial Africa* speaks volumes. In 1886, Henry Morton Stanley led a disastrous, arrogant, ill-advised expedition up the Congo River to "rescue" Emin Pasha, a German naturalist and physician who had been appointed governor of Equatoria — and who in fact needed no rescuing. Half of Stanley's 389 native porters died. Among those that survived, according to Hoschschild, "When they ran out of food, they caught and roasted ants." Ants, then, were seen as a food of desperation, to be eaten when they ran out of "real" food. But even what we eat when desperate is constrained by what our cultures have conditioned us to see as possibilities for food. The porters saw the ants as food; their American scribe did not. I am reminded of a saying among the Yansi, an indigenous group in Zaire: "As food, caterpillars are regulars in the village but meat is a stranger." Which says as much about what they consider to be "meat" as it does about the importance of caterpillars in their traditional diet.

As I've been writing this chapter, I've been reading my grandfather's diary from the 1920s famine in the Ukraine, in which insects aren't even mentioned as a food of desperation. At one point he writes, "And what did we eat? Rats, dogs, crows, horsemeat, bread made from pumpkins, beets, millet porridge and millet chaff." Like most people of northern European descent, he was aware of insects, but when thinking about food, they were visible only as pests. My grandfather was not unusual in his views. These deeply held attitudes about what can be used as food, and what cannot be imagined as food, are difficult challenges for those promoting entomophagy as a response to a globally desperate food security situation.

Even among those who are earnestly trying to be sensitive to indigenous knowledge and culture, as well as ecological resilience, we encounter a subtle deference to the language of colonial culture. Mopane worms, caterpillars of the emperor moth *Gonimbrasia belina*, have been integral to the food cultures of many tribal groups in southern Africa since prehistoric times. Nevertheless, recent published reviews of this important food source refer to the practice of eating mopane worms as a "livelihood strategy" for marginalized households whose "livelihood alternatives" are otherwise limited. *Livelihood strategies* is a phrase often used by development specialists to describe how people in economically and politically disempowered communities cobble together a way of life, and the term encompasses provision of food and shelter, income-generation, and daily living practices. As sustainable development became a catchword globally in the 1990s, the phrase *sustainable livelihoods* was increasingly used in development aid circles. The International Fund for Agricultural Development refers to the Sustainable Livelihoods Approach (often shortened to SLA, but not to be confused — by those of us of a certain age — with those Patty Hearst kidnappers, the Symbionese Liberation Army)[80] as "a way to improve understanding of the livelihoods of poor people."

I have no objection to the use of SLA terminology per se, but I have yet to hear it applied to urban professionals and Silicon Valley acolytes in cities of North America or Europe. Yet one could argue that it is these livelihoods that are unsustainable, not those of villagers in Malawi. This reminds me of the OIE program to assess animal health infrastructure. When one Chief Veterinary Officer suggested that the countries of Western Europe or North America should undergo such reviews, the OIE officials were offended. This was a tool for evaluating "developing" countries, not "us."

Is entomophagy a kind of neocolonialism, or is it, as I am hoping, a way for modern urbanites to develop SLAs? In this century, a series of FAO workshops brought what had been a niche activity in the world food debates out into the open. I've already referred to the reports from these workshops — *Forest Insects as Food: Humans Bite Back* (2010) and *Edible Insects:*

*Future Prospects for Food and Feed Security* (2013) — that launched a kind of global beetle-mania. The reports are full of surveys and case reports of "ethnic groups" (aren't we all, in some way, ethnic?) who ate, and still eat, insects in Africa, Asia, and Latin America. Suddenly entomologists, archeologists, and anthropologists were combing the world, looking for more reports of people who ate, or had eaten, bugs. Behind all this were the billions of people who, until they were "discovered" by the new entomophagists, had been quietly, with little fuss, going about eating locusts and weevils and termites just as if they were regular food. The stories of these people are, quite literally, all over the map, and are as culturally diverse as the ecosystems within which they emerged.

As insects creep into our cooking repertoire, we would do well to pay attention to those stories — which insects were eaten, and why, and how were they prepared? At its best, the path to normalizing insects on the plate is a path that leads through mutual respect to greater ecological and cultural understanding.

In parts of southern Africa, people eat the stink bug *Encosternum delegorguei* when the mopane caterpillars are pupating underground and unavailable. But stink bugs are not just eaten "as is"; to make them palatable, they are washed three times in warm water, then boiled, and then sun dried. To detoxify an armoured ground cricket (*Acanthoplus spiseri* ) before eating, one should pull off its head, remove its gut, boil it for at least five hours, and then fry it in oil. An insectivorous adventurer who ignores this advice may end up being seriously inconvenienced by an inflamed bladder. Mopane worms need degutting, and dung beetles cleaning, before they are eaten. We can improvise based on the original recipes, but the traditional entomophagists among the minority peoples of China and Africa and Latin America should be recognized as the Julia Childs of the new entomophagy movement.

In many cases, Julia's children may be already gone. I suspect it's too late, for instance, to find American indigenous cooks who can show us how to best prepare a desert fruitcake or a locust soup.

Some recipes, however, live on in current practice. Among the insects that have been traditional sources of food for people in various parts of the world, those that have managed to find the cushiest jobs in the postmodern economy are palm weevils, mealworms, and crickets.

I've already given mealworms the once-over in connection to nutrient content and animal feeds, and I'll come back to crickets shortly.

For those of us who grew up in temperate zones, the least familiar of the three are the tropical palm weevils, whose natural ecological niche I described earlier. Originating from tropical Asia, palm weevils are often considered pests because they can transmit parasites from tree to tree. They are also a culinary delicacy in much of the non-European world. Semi-cultivation of palm weevils, in which the farmer-foragers create larvae beds by knocking down trees and exposing the pith for the weevils to lay eggs, has been reported from Paraguay, Colombia, Venezuela, Papua New Guinea, and Thailand. The Jotï, a semi-nomadic people from the Venezuelan Amazon, cultivate two species, *Rhynchophorus palmarum* and *Rhinostomus barbirostris*, although they prefer the latter for its richer flavor. In Southeast Asia, where the larvae are known as sago worms, they are called "sago delight" when fried. A Cameroonian cookbook describes red palm weevil larvae, also known as coconut larvae, as "a favourite dish offered only to good friends."

In Thailand, demand for sago larvae, once foraged for occasional snacks or eaten as a form of pest control, has taken off to the point that farmers cannot keep up with the increasing demand. The traditional farming methods are to cut down cabbage or sago palm trees, drill holes into them, and then place breeding pairs of weevils next to the holes. The dynamics between increased demand and traditional foraging are a worldwide phenomenon, and increases in human populations and the felling of palms for farming are already being felt ecologically and culturally. In Venezuela, the Jotï now walk four to twenty hours farther than they once did to find palms to prepare for weevil cultivation. In Thailand, traditional methods have been replaced by putting breeding pairs into plastic containers, where they are fed ground palm and pig feed. By 2011, 120 Thai farmers were producing forty-three tons of weevil meat annually, as well as frass, which was used for fertilizer.

In Ghana, to avoid destructive overforaging, the Aspire Food Group are feeding their palm weevil larvae a mix of old, rotten palm trees and palm wine. Aspire was founded in 2013, when five MBA students at McGill University, none of whom had background in farming or insects, won the million-dollar Hult Prize, "a start-up accelerator for budding young social entrepreneurs." Their plan was to work primarily in Ghana, the United States, and Mexico. On their website, they announced that their mission was "to provide economically challenged, malnourished populations with high protein and micronutrient-rich food solutions derived from the supply and development of insects and insect-based products."[81]

When Shobhita Soor, one of the founders of Aspire, first told me over coffee at a Montreal café about the nutritional marvels of palm weevil larvae and Aspire's Ghana project, I was skeptical. What could a small group of MBA students with more experience in spreadsheets and texting than in farming, international development, or insects, hope to accomplish? Was this just another example of neocolonialism with wealthy, well-meaning professionals from Europe and North America trying to tell Ghanians how to live sustainably, even though we have not proven ourselves able to do so? In January 2016, when the *Guardian* newspaper reported on the project, I was ready to start shaking my head in cynical sadness. A quote from a Ghanian professor about the possibility of reviving the lost practice of insect-eating in Ghana among the urbanized middle classes, however, gave me second thoughts, as did Soor's emphasis on developing sustainable farming and business practices. "We are not here to change the way people eat or tell them what to eat," she was quoted as saying. "We are here to provide a desired source of protein and iron in a much more accessible way. Palm weevil is a great source of iron and protein."[82] Maybe the *Guardian* just has skilled journalists (it does), and Soor is a deft PR person (no doubt), but I couldn't help thinking that maybe a group of idealistic, enthusiastic young people could do what the mob of old, rich, white men who doled out the money never could.

Beyond palm weevil larvae, we can cite the importance of mopane worms in eastern and southern Africa, which have valuable ecological work to do *and* are nutritionally rich food supplements *and* provide substantial cash income for rural households (25 percent in some parts of southern Zimbabwe). African countries have provided an endless stream of entomophagy stories and research reports. Almost all the people living in the forests of the Central African Republic are reported to rely on insects for their protein. A study of the indigenous Gbaya people in the Democratic Republic of Congo (DRC) found that insects formed 15 percent of their protein intake. Another study announced that "the average household" in Kinshasa, DRC, was eating 300 grams of caterpillars a week.

Termites, especially the fungus-farming termites of the genus *Macrotermes*, are a desirable food item across sub-Saharan Africa. A 2013 review of termites as food (which wasn't included in the systematic reviews I cited earlier) asserted that they "contain significant proportions of proteins, fats and minerals. The oil is of high quality with significant amounts of polyunsaturated omega-3 fatty acids. The termites have unique nutritional qualities that can be exploited to provide high-quality diets, especially in the developing countries, which have been plagued by iron and zinc deficiencies as well as poor supply of dietary polyunsaturated fatty acid sources."[83]

People who eat termites sometimes also eat the clay from which the termite mounds are built, which is apparently high in kaolin, a treatment for stomach upsets. In ostensibly more sophisticated Western countries, some of us grew up eating Kaopectate, which, at least in its original formulation, was much the same thing — to stop diarrhea. Pregnant and breast-feeding women are said to benefit especially from eating termite mound clay; the benefits, according to one research report, are improvements in calcium intake, stronger fetal skeletons, and increased birth weight, as well as reduced hypertension associated with pregnancy.

Cone-headed grasshoppers (*Ruspolia nitidula*) are also popular in Uganda and have been, periodically, a significant source of income, their price per unit weight exceeding that of beef in the markets. This was especially true in the 1990s, when plunging coffee prices deprived many Ghanians of their primary source of cash income. The reported constraints were that the grasshoppers had a short shelf life, and that, when retrieved from the drums where they were stored, they would bite.

It is all very well to talk about eating grasshoppers, stink bugs, and bamboo worms in Africa and Asia, but it will take more than just a leap of the imagination for bugs to go from there to grocery stores in Canada and the United States. One of the recurring questions among the new promoters of entomophagy is how insect-eating can be taken from its remnant populations of disappearing traditional, ecologically grounded communities and integrated into an emerging, Western-dominated, global food culture.

The success story most often cited of a food once marginal and now mainstream is that of sushi. If Europeans and North Americans can, in one generation, learn to eat raw fish as prepared in Japan, then why not insects? There are even evolutionary arguments for pursuing this line of marketing. According to the latest genetic research, insects may be thought of as descendants of terrestrial crustaceans; their closest living relatives are blind, cave-dwelling remipedes. From an entomophagical point of view, perhaps this "sisterhood" of insects and crustaceans can be used to alter the public imagination. With all due respect for, and apologies to, Daniella Martin's YouTube scorpion-eating performances, distinguishing and imaginatively distancing edible land crustaceans from spiders and scorpions might offer more effective marketing possibilities.

Japan was the source of the American and European infatuation with sushi, and now is also on the forefront of the new entomophagy movement. Japanese culture has a long history of explicit engagement with insects as entertainment, natural phenomena, and food. The Japanese word *mushi*, which can refer to a bug, a germ, an insect, or a spirit, reflects this complex cultural perspective.

This relationship with insects is changing rapidly, as a result of both Western interests in entomophagy and how that interest interacts with the modernization of traditional cultures in Japan. Some practices, such as foraging for caterpillars while collecting firewood, have largely disappeared because not many people use firewood for cooking anymore. Apparently Emperor Hirohito had a fondness for fried wasps, seasoned with soy and sugar and eaten with boiled rice, but imperial Japan is not generally held up as inspiration for the future world. As in China and Korea, silkworm pupae, a by-product of silk production, are still widely consumed in Japan. Since they were domesticated in China some five thousand years ago, the cocoon size, growth rate, and FCR of silkmoths (*Bombyx mori*), which are the basis of 90 percent of the world's silk production, have all improved relative to wild forms. About a million people are employed in the Chinese silk industry, which annually produces more than 140,000 tons of silk, 800,000 tons of fresh cocoons and, more to the point of our discussion here, more than 400,000 tons of dry silkworm pupae. The spent pupae have multiple uses, including as fertilizer, human food, and animal feed. White mulberry trees (*Morus alba*) are the preferred food of *Bombyx mori*; they are native to northern China but have now been cultivated and naturalized in many other countries, including Japan.

Eating grasshoppers, once widely practiced in Japan, is now on the decline. Hunting and eating of wasps, giant hornets, and yellow jackets has been reported in documentary films, academic reports, and public media, but a declining supply of some of these is leading to imports from Korea and China.

It is not yet clear which bugs will find a home in the brave new entomophagical world, or into whose kitchens they will fly. From what I have seen and read and heard, one man is ahead of this wave of insect-eating culinary changes. Born into a community in central Japan where insect-eating was normal, and now based in that most futuristic of cities, Tokyo, Shoichi Uchiyama has become a national and global champion of insect-eating. The author of several books on the delights of entomophagy, he leads and inspires the Konchu Ryori Kenkyukai (Insect Cuisine Research Association). Daniella Martin describes her memorable encounter with him in her book *Edible*. Having watched Uchiyama-san[84] on YouTube and seen him celebrated in popular newspaper reports and an NKH World documentary on insect-eating in Japan, I had to visit Uchiyama-san as well.

Together with the brilliant and well-organized Yukiko Kurioka of Japan Uni Agency (who had negotiated the Japanese publication of my previous book, *The Origin of Feces*) Uchiyama-san had arranged for me to do a book reading and give a talk the day I arrived. The reading was at a bookstore that took up the entire very large second floor of a department store. There were a lot of people in the store and, even if they were looking for graphic novels and manga, I was impressed. The topic of my talk was "Mimicking Nature: Entomophagy and Feces in a Sustainable Society." Karen Kawabata, the daughter of my friend and colleague Zen Kawabata, translated for me.

The audience was, if not rapt, then at least politely attentive, and even laughed at some of my attempts at "global" humor. What's more global than Walt Disney, for instance, and the thought of Bambi's mother eating her newborn's feces? In the Q and A, one guy wanted to know if he should shit in his garden to improve the soil. I advised against it. Another guy suggested that Europeans didn't eat insects because the Bible said not to. I said I thought it had a lot more to do with climate and landscape than religion.

I had given Uchiyama-san a bag of Moroccan-spiced crickets from Entomo Farms in Canada, and several plates were passed around for people to sample. During the signing, a woman pressed a magazine into my hand, opened at a particular page that seemed to show young women cooking insects. It was only after I got home that I opened it and realized it was not a Japanese version of *Bon Appétit* or something similar. In an email to me, Yukiko explained: "*Friday* is a weekly magazine which mainly covers celebrities and their scandals as well as cultural trends. The sticky note seems to be addressed to somebody else. It says 'I'm sorry that I did not send this copy earlier. Sega (a person's surname).' The article is about the young girls eating insects. They drink the vodka in which belostomatidaes [giant water bugs, Hemipterans to entomologists, toe-biters to those inclined to entomophobia] are pickled and eat fried locusts and larvae of butterflies as nibbles. The article mentions that FAO recommends entomophagy. The journalist joined the event by Mr. Uchiyama on May 18. The article goes on to explain different dishes and the charms of insect eating."

The charms of insect-eating, indeed. And drinking vodka. And of photoshopped Japanese girls in their panties. But I'm

sure people only buy the magazine for the articles on entomophagy. In any case, I suppose what this illustrates is one of
the ways that eating insects is already insinuating its way into
urban popular cultures.

The next morning, Kyoko and Kenichirò Iizuka, my local
guides, took me out to join Uchiyama-san and several other
people for insect hunting and roasting along the Tama River.
After a half-hour train trip out of the city, we walked a few blocks
to a small shop, where Uchiyama-san picked up his bicycle. The
bike was loaded with the accoutrements and equipment necessary for hunting and picnicking, including a tarp, butterfly nets,
a propane camping stove, and bags of cicada (locally caught)
and ant larvae (imported from China). Our intrepid hunting
group walked in the baking heat to the riverbank, where we
parked under a bridge. Then we set off to catch lunch.

As I walked down the narrow path among the tall grasses
and shrubs, I heard a kazoo-like buzzing nearby. I approached
the musical branch and was the reluctant witness to a thumb-
sized hornet devouring a smaller, green insect, probably a grasshopper or mantis. Having heard all the horror stories about
Asian killer hornets, I slowly backed away. Later, overcoming
my better instincts for the sake of science, I checked into a much
louder flapping in a tree. A praying mantis — several centimeters long — had her teeth into a cicada. When I netted the pair
and showed it to Uchiyama-san, he informed me that the mantis
was pregnant (and thus eating for more than one!). The Japanese
name for the mantis is *kama kiri* which translates literally as
"sickle cut" (*kiri* meaning "cut," as in *hara kiri*).

At first, a lot of grasshoppers and mantises whizzed away
just as my net came down. I finally caught a grasshopper, but

when I swooped my net down on a second one and tried to stuff it into my ziplock bag, the first escaped. Eventually I learned how to cultivate patience, waiting until a bug settled before moving with startling speed to catch it, and then shaking my previous captives down to the bottom of the bag before stuffing in the newcomer.

Back at camp, under the cement bridge, Uchiyama-san and his helpers had set up the stoves and frying pans and busily cooked up the various bugs and larvae. I was told the flavor was "nutty," and I suppose it was. I wondered what kind of nuts, and a couple of us decided that maybe the pan-fried cicada larvae tasted a bit like almonds. Now, when someone asks me what almonds taste like, I can say, "a bit like fried cicada larvae."

After this adventure, Uchiyama-san and his coterie went back to his home to clean up. We met again at a railway station a few stops away, and then we were off to Akihabara, the world center of big-eyed anime cartoons, video games, movies, and manga. Japan International Volunteer Center (JIVC), which was running a community-based project in Lao PDR on the relationship between insect foraging and forest conservation, is located in an alley just at the edge of Akihabara. As I expected, results from the JIVC project looked promising but uncertain, as tends to be the case where local people are given hunting or foraging rights in protected areas. The theory was that if indigenous people were given these rights, they would protect the resources. But such a strategy does not account for political manipulation and the massive financial pressures that can occur if those resources gain traction on the open market. I spoke briefly about excrement, and the VWB/VSF cricket farming project in Laos. Uchiyama-san and his staff prepared snacks of saltine crackers smeared with what I think was some kind of insect pâté, and topped with either crushed crickets, ants, or cicadas.

The next afternoon, Kyoko and Ken met me in the lobby of my hotel and we headed over to where Uchiyama-san has his office for an afternoon of insect cooking and tasting. There were about a dozen people, including a grad student in ESL from Ohio and an editor from the publisher Tsukiji Shokan. The space was cramped and stacked with books. According to Ken, Uchiyama-san works for a publisher that specializes in Russian literature. But one of the few English books on the shelf was by Allan Ginsberg. Suddenly it all made sense! (Well, as much as Ginsberg ever makes sense.) Are the beat poets a gateway into the mainstream? If Uchiyama-san and le Festin Nu are leaders, then perhaps the answer is yes.

The menu included hornet larvae, silkworm pupae, and silkworms. The geographic origins of these insects were not always clear to me. Some, I think, must have been imported. The silkworm pupae were white and pink and yellow. Apparently silk producers have bred various colored strains. We snipped off the ends and the larvae dropped out. Zen roasted them in a small pan over a camp stove in the street to get the "chaff" off. The hornets were bought from a company that cleans hornet nests from people's houses, so it was doubly virtuous to eat them. We made tea from the feces of worms that had fed on cherry blossoms. The tea was cherry-scented and, if you didn't know where it came from, light and tasty. We also tried green tea made from silkworm larvae poop, which tasted like green tea made from silkworm larvae poop. One of Uchiyama-san's assistants made noodles from buckwheat dough that included powdered whole bees.

Reflecting on this later, I tried to discern where the path might be from this type of street cooking to North American kitchens and restaurants. There did not seem to be an obvious route.

The second part of my agenda in Japan was to go hornet hunting near Nagoya. Yukiko met me at the hotel at 7:30 and guided me to the bullet train. She handed me a train ticket and a schedule with pictures of the gates I was to pass through, and pointed me in the right direction. The Japanese trains are exactly on time, and smooth, especially that bullet train. At Nagoya station I had to rush to make the connection to the fast train to Ena. In Ena, I caught the one-car train that reminded me of the hand-painted VW bus my future wife drove back in the 60s. We chugged up through tunnels and deep green chasms into the mountains to Akechi, where I was met by Shoko and her two year-old daughter, Soyoka. On the way out of town, we stopped at a small grocery store, and I noted jars of what looked like pickled hornets on the shelf. We then drove out to Kushihara, a village of fewer than a thousand people, where their AirBNB, named Lumberjack, is located. Shoko's husband Daesuke-san, the real-life lumberjack, who does indeed own and run a small sawmill, had assured me in an email that we would go "bee hunting."

Researcher Charlotte Payne, who has spent considerable time studying in the area and published some essential-reading academic papers on entomophagy, assured me that he meant hornets. She also explained that although the people in the region did sometimes hunt giant Asian hornets, at the time of year I was arriving, the animals we were after would be "kurosuzume-bachi" — *Vespula shidai* or *Vespula flaviceps* — generally referred to in English as wasps.

When I inquired later about this hornet–wasp confusion, Charlotte explained that "the confusion about terminology comes from a quirk of the Japanese language: the word 'hachi' ('bat') occurs in the common names for all Hymenoptera except ants — so, hornets, bees and wasps can be referred to by this term. Since 'bee' is the first Hymenoptera species that most children learn about, most people translate 'hachi' as 'bee.' Hence, many people interested in entomophagy leave Japan thinking that they have eaten bee larvae (often marketed to English speakers as bee babies) when in fact they've had the larvae of *Vespula flaviceps/shidai*, which are the most commonly eaten wasp species in the country. (Other communities allegedly eat other orders of wasps though I have never seen this first hand.)"

The next morning, after the usual breakfast of rice, fermented beans, fermented pickles of eggplant and cucumbers, and miso soup, we were off to hunt wild hornets (this adjective probably being unnecessary as there are no domesticated hornets or wasps). I climbed into Daesuke-san's pickup and, followed by one of his sawmill workers in another small car, drove about half an hour along a single-lane, paved valley road between cedar and Japanese cypress forests. Arriving at a cluster of houses beside the road, we met seventy-six-year-old Haru-O, otherwise called Haru-san, the expert hornet hunter. A short, weather-worn man in a baseball cap, jeans, and boots with the toes separated — which (excepting me) seemed to be the uniform of the day — he had been doing this for fifty years. Also there was a seventy-one-year-old community leader of unknown (to me) status.

We drove in tandem up a sometimes muddy, sometimes gravelly road between steep hills, and finally stopped near a large grader, where the road was going to be extended. Haru-san prepared sticks with a spit at one end, each of which he pressed a strip of squid into (some people use eel, which is what Daesuke-san called it at first, but which his wife, Shoko, later translated as squid); these sticks, each marked with a pink ribbon, were stuck into the roadside at various intervals. Then, we waited.

In the meantime, Daesuke-san showed me a plastic box that held thin white threads that seemed to broaden at one end (a bit like flossing strips). When, finally, one of us saw a small black hornet chewing on the bait, Haru-san took a small bit of squid, worked it into a pearl-sized spit-ball, and attached the string to it. He then approached the hornet, and prodded it to take the squid-ball from his hand. The hornet took a bite and zipped away, trailing the white thread. We watched it disappear into the green foliage between the cedars, and Daesuke-san clambered up the steep hill through wet, partially decayed logs, ferns, scree, and tree trash until he lost track of it. We waited for another and repeated the process. This time he was able to follow it farther. After about three baited hornets, we were able to find a few holes set back into the earth in the shadow of a rotten log and a curtain of ferns. While three of us waited next to the nest, Daesuke-san went back down to help Haru. For a while, we watched hornets coming and going from the holes. While I nervously stepped back, the seventy-one-year-old whacked the earth a few times to watch the hornets come hurrying out to see who the intruder was.

Finally, Haru-san and Daesuke-san climbed up, bringing a couple of twelve-inch-square wooden boxes. Haru-san pulled on a beekeeper's hat and veil, and thick gloves, then dug into the earth below the holes. In about five minutes, he had unearthed a large grapefruit-sized hornet's nest from the wet red humus and dirt. He plunked it into one of the boxes. The box was too small, but while Daesuke-san clambered back down to get a larger box from the truck, Haru-san kept digging and dropping handfuls of hornets and nest into the box. When Daesuke-san arrived with the bigger box, Haru-san transferred the nest to the new box and added more hornets, including the queen, who promptly crawled down into the nest. He tied the lid on the box and we all returned to the vehicles.

Back at his house, Haru-san showed us the twenty nest boxes he owned. He fed them chicken liver (on a string hanging outside each front door) and rock sugar. He would feed the nests for a few months, until the November festival, at which time (so I understood) there would be a contest to see who had the biggest nest. Then they would eat most of them and let some go.

On my final morning in Japan, during my usual breakfast (no insects), Daesuke-san showed the NHK World video about Uchiyama-san and insect-eating in Japan to two of the workers from his sawmill. The younger guy, probably in his twenties, grimaced. After the film was done, and the guys had gone back to work, Daesuke-san came over and sat next to me at the wooden table. He seemed agitated, as if he had something important to say but wasn't quite sure how to say it.

He was concerned, he finally said. With all the film crews and the researchers coming into his village, he was upset that all these foreigners were concentrating only on the fact that they sometimes ate insects. Sure they did. But that was not who they were. They were a diverse and multifaceted community, living regular lives, trying to be sustainable, to make a living, to have meaningful lives. They sometimes ate insects, but they were not defined by this. He wanted me to remember that as I worked on my book.

Waiting for my delayed flight at Tokyo's Haneda Airport, I thought more about what Daesuke-san had said. It was true. I sometimes ate locally foraged maple syrup, and apple butter made from locally grown apples, but those foods did not define me. I suppose the foods I grew up eating — *verenicke* and *borscht* and *porʒeltche* and *pfeffernuesse* and *paska* — were, by virtue of their historical and familial baggage, part of my sense of identity. Throwing a handful of bugs into the *borscht* and adding cricket protein powder to my *paska* would change their ecological and nutritional values, but not how they related to my sense of cultural identity. Food and taste are very deeply rooted in the warp and weft of interwoven personal, family, and ecological histories. If we changed the content of our food but kept the form — substituted bugs for ham in the bean soup, for instance — would that, across generations, also change the culture that gave us meaning, the sense of who we were? I suspected so, but wasn't yet sure how.

In March 2004, I had been at the Great Library of Alexandria in Egypt. The occasion was a special conference of that massive global initiative of the early 2000s, the Millennium Ecosystem Assessment. The conference was titled Bridging Scales and Epistemologies, and it emphasized the challenges of listening across cultures, having mutually respectful communications across different worldviews, and creating effective linkages that could encompass individuals, villages, governments, and, indeed, the whole earth. If the world of insect-eating is any indication, it seems to me that the important conversations needed to begin designing these bridges have barely begun.

SHE CAME IN THROUGH
THE KITCHEN WINDOW

*Culinary Renewal from the Margins*

• • • • • • •

*Cool stink bug cream and a nice weevil tart,*
*I feel your taste all the time we're apart*

Southeast Asia, as many YouTube videos and enthusiastic travelers will tell you, offers a veritable buffet of insectivorism and seems to be at the forefront of linking the cuisines of insect-eating and non-insect-eating cultures.

Introduced into Thailand in 1988 based on technology developed by Khon Kaen University, cricket farming and demands for its products in Thailand and beyond have exceeded even the most optimistic expectations. By 2011, twenty thousand Thai cricket farmers were putting out more than seven tons annually. Schools had introduced programs to fortify school lunches. The highest input costs were feed-related. Many were giving the crickets chicken feed, although vegetables such as pumpkins,

194

cassava, morning glory leaves, and watermelon were often fed just before harvest to improve taste.

In neighboring Lao PDR, cricket farming has been slower to commercialize, and has only been around since the early aughts. Nevertheless, various ethnic groups in the region have been eating insects for a long time. Even now, although under duress to give up their "primitive" habits by television-inspired globalization, some 95 percent of Laotians are reported to eat insects of one sort or another. The surviving entomophagical practices include, for people living near water such as a paddy field, scavenging for diving beetles, water scavenger beetles, water scorpions, giant water bugs, stink bugs, and dragonfly larvae. In the dry season, people eat the larvae and adults of dung beetles. Stink bugs are sold fried or live to be prepared at home. Cooks remove the stink by boiling them, and then make them into a paste to be eaten as a side dish. People in South Africa, Malawi, Papua New Guinea, and Mexico also eat stink bugs.

In August of 2015, Thomas Weigel (the cricket project manager for VWB/VSF) took me to the Dong Makkhai market, just outside of Vientiane, where the vendors sell a variety of products foraged from Laotian forests. One of the authors included in *Forest Insects as Food: Humans Bite Back* reported that "among the edible insects [at the Dong Makkhai Market] the biggest sellers are weaver ant eggs (23 percent), grasshoppers (23 percent), crickets (13 percent), honeycombs (13 percent), wasps (9 percent), cicadas (5 percent) and honey bees (5 percent). The highest price is paid for young cicadas — about US$25/kilogram."

The foods Thomas and I saw on display in the market were as stunningly diverse as the reports indicated. Heaped on the display tables were mushrooms, seeds, frogs, turtles, honey laced with some kind of bitter, medicinal plant product (passably okay, I discovered later, with Nutella), and a wide variety of six-legged wildlife — different sizes of grasshoppers, small white crickets, house crickets, mole crickets, wasps, wasp eggs and larvae, dragonflies, and dung beetles. Many of these were already roasted or fried, but in a few cases live crickets were on display, freshness assured by their scrabbling attempts to escape their deep plastic bowls, resident flies circling the rims. Thomas and I bought a couple of portions of fried crickets and some larvae of unknown (to me) provenance. We crunched these as we wandered around looking at other freshly foraged offerings. At one stall, where we stopped for lunch (noodle soup, no insects), several young women (and their mothers) offered themselves to us as suitable marriage partners.

At another stall, we were offered more pedestrian (compared to the marriage proposals) fresh, fat, white hornet larvae. Here we accepted the offer; the soft, velvety larvae were smooth on the tongue and of a pleasantly cool, buttery, custard-like consistency. On the table, Asian giant hornets, fat and a few centimeters long, wandered lost and disoriented over the cells of the broken comb-like nursery of their dead brood. It made me sad to see them, but thinking about food often gives me a sense of my own mortality and of being inextricably caught, squirming, a turmoil of conflicted feelings, in the web of life, with its expected death, unexpected beauty, and necessary loss. In deference to the many indigenous peoples of South America, Mexico, Africa, Australia, and Asia who consider bee, wasp, and hornet brood delicacies, I bought a piece of the nest with larvae and some dead giant hornets, which the seller mixed with kaffir lime leaves.[85]

Back at the house, I improvised a curry with the hornets and their babies and decided that if this was going to work for me, first, I needed a better recipe, and second, I should probably consult with a chef from one of those many indigenous cultures.

A few days later, several of us — including Fongsamouth Southammavong, Vice Dean of the Faculty of Agriculture, and Daovy Kongmanila, an animal scientist at the National University of Laos— drove through the flat, green landscape to visit the VWB/VSF cricket-rearing projects. One of these, near Vientiane, involved sixteen farmers in the village of Hatviangkham (HVK), near the university. The aim of this was to create value-added products for crickets to be marketed in Vientiane. The other, in Bolikhamxay district, an impoverished area that had recently suffered serious flooding and food shortages, was about three and a half hours' drive south of the capital. If I had come just a week earlier, the roads would have been completely impassable. Indeed, while we were in Bolikhamxay, we heard that a dead body had been found — presumably drowned in the floods. Even now there were large areas of red, slimy ruts.

After having taken a design workshop, the farmers had built their own cricket-rearing crates, about a meter by two meters and one and a half meters high. The crickets scrambled around, hiding as I watched, retreating to large paper maché egg cartons that were propped against the walls of the pen. Farmers left out trays filled with sawdust or rice bran where the crickets could lay their eggs. The farmers (all women) were feeding chicken feed to the crickets, occasionally supplemented with fresh produce such as water spinach and cassava leaves. For the last five days before harvest, the crickets were on a vegan diet. After forty-five to fifty days, just after eggs had been laid for next cycle, the farmers said they harvested about five kilograms of crickets. They knew when to harvest, they said, because at breeding time the males chirped like mad and scrambled after females. Then they waited until the females laid their eggs. The frass was used for garden fertilizer.

Our lunch was courtesy of Fongsamouth's college roommate, now a director in the Provincial Agriculture and Forestry Department. Our four-wheel-drive vehicle bumped and slid along a slimy, winding red mud road between hills and alongside a man-made lake. At road's end, we walked to a gazebo overlooking the water. I leaned on the railing as the director's staff fired up the barbecues. An electric line was strung down from the hillside to a lightbulb dangling over the water. This would be turned on at night, the director said, to attract insects for the carnivorous fish in the lake; the fish were caught for food. Cassava leaves were thrown into the water to supplement the diet of the herbivorous fish. A wooden house up on stilts perched on the other side of a small earthen dam. Next to it was an insect trap, similar to the ones I had seen in Cambodia, consisting of an upright panel of metal corrugated roofing with a fluorescent tube light stuck vertically in the center of it and a tub of water below. I was told that this is the preferred method for catching insects in Laos and is used by many households. I walked over and looked into the water trap: a few giant water bugs, dragonflies, moths, and an iridescent jewel beetle. Andrew Vickerson at Enterra Feed in Canada, who had worked in Cambodia, had worried aloud to me about the problem of "by-catch" with these systems. I supposed this might be a problem for any scaled-up version of insect foraging.

Daovy explained to me that many people used the bugs

caught in this way for themselves or to feed fish or other animals. She related how one morning she and her husband woke up later than usual, and their chickens had eaten all the bugs in the traps next to their house.

The other VWB/VSF project involved both cricket farming and the production of chips and salsa, both with crickets as ingredients. The chips were made of manioc flour, dried and ground garlic and shallots, and dried crickets. The manioc dough was rolled into sausage-like shapes, wrapped in plastic, and then steamed. The steamed dough was cooled in the refrigerator overnight and then sliced thinly. The chips were fried in hot oil. Before the final production stages, Thomas and his colleagues had visited pubs and beer shops (it's a living, eh?), distributing samples to elicit orders. By the time the orders were finalized, however, some of the HVK farmers had eaten their crickets or sold them in local markets. To offset this, Thomas needed to pick up a half kilogram of crickets. So, on the way into the city after the visit to Bolikhamxay, we stopped at a commercial farmer unconnected to any foreign aid or development project to see if we could buy from him.

The farmer was a young guy, in his twenties; he had seven pens of crickets and was growing them on chicken feed. His production — eight to fifteen kilograms every six weeks, selling at 35,000 LAK[86] per kilogram to market sellers (who resold it at 50,000 LAK/kilogram) — was good. We asked him how he had learned cricket farming. From the internet, he replied, and from other farmers. Now he was at a stage where he shared eggs with several other people in the area who were interested. As we were speaking, another guy strolled over from a nearby building. He explained that he was a section head in the Department of Agricultural Extension and Cooperatives, Ministry of Agriculture and Forestry, and that he was growing crickets as a hobby. He had six crates. I wondered where development projects fit into this context. Were they unfair competition? Were foreigners once again stepping in thinking that they knew better? Or (my preferred interpretation) were the projects a stimulus to promote new farming and food practices?

The following day, at the university food science laboratory, I watched as several of the staff prepared five versions of cricket salsa for consumer testing. They worked with mortars and pestles in the lab and a small outdoor barbecue for roasting onions and garlic. After preparing the salsa, they tried the different versions out on individuals on or near the university campus. Later they returned to the laboratory and entered the information they'd collected — about color, smell, and taste, for example — into a spreadsheet. The "best" salsa would then be presented to the HVK villagers, who would (based on previous experience) modify the recipes as they saw fit, changing ingredients "more or less to taste." This, it seemed to me, was a way for the villagers to claim both ownership and agency over the process. These modifications also, for better or worse, created challenges for anyone attempting to industrialize the process.

In a 2015 article in *Trends in Food Science and Technology*, Matan Shelomi, a postdoctoral research fellow at Max Planck Institute for Chemical Ecology and a specialist in the evolutionary genetics of insect digestive enzymes, suggests that we should think about insects in Western and European food systems the way we think about nuts; we already describe insects as having a nutty flavor, and many of the more popular ones, such as mealworms and crickets, are already used the way nuts are

— as optional ingredients in otherwise complete foods, as snacks, or as a source of cooking oil. Indeed, fried up with some salt, garlic, and kaffir lime leaves, heaped over fresh-cut cucumber, the karaoke crickets Thomas and I had shared had tasted fresh and crunchy, like beer nuts for drinkers with nut allergies. The analogy is useful but, like all analogies, imperfect. In terms of spoilage and food safety, insects are still more like shrimp.

Cricket farming in Lao PDR and neighboring countries involved small initial capital investment and appeared to be a good way for poor people living on the fringes of large urban centers to make a living and improve the nutrition of their families. The logic of international development suggested that the next step was to "scale up," and Thomas was exploring the idea of working with a national beer-brewing company to use their brewer's waste as cricket food, which would be a way to improve growth, reduce costs, and close a frayed and broken ecological connection. This all sounds good, but I get nervous about luring farmers into the industrial global agri-food system so they can make more money. For one thing, if backyard chicken rearing is any indication, as soon as the crickets go from being subsistence to commercial, the management and income will shift from women to men, and one of the original rationales for this work will be lost. I also worry that, like Thai people who flocked from the countryside to Bangkok in the 1960s, rural Laotians will use their increased disposable income to build giant shopping malls and to sell enough cars to turn the streets into noisily impassable canyons between skyscrapers. Already Thomas could point out to me the Chinese-financed apartment blocks and shopping malls bulldozing their way across the green landscape on the edge of Vientiane. Was this the future? Would there be a place for small-scale cricket farmers just trying to keep their families healthy and well fed? Was it any of my business what they did with their increased income?

Cricket farming might improve household nutrition and increase income for smallholder farmers in Southeast Asia, and improve the lot of rural women there, but what roles could it have in North America, where industrialized livestock systems were well entrenched?

Canada's Entomo Farms offers an example of what might be done. Rather than starting from the wants of consumers, Entomo Farms started from the farm and decided to take a more direct approach to putting insects on the dinner table. Ecologists Tim Allen and his colleagues Joe Tainter and Tom Hoekstra have called this approach — starting from the resource base rather than the consumer — "supply-side sustainability." Marketing themselves as "the future of food," Entomo declares that they are producing the "world's most sustainable superfood." Having started out as "reptile feeders," producing insects for reptiles and fish, they had made the cross-species leap just a year before my 2015 visit. Dedicated, enthusiastic, smart, and media-savvy, the owners have been profiled in Canadian newspapers and radio programs and celebrated their innovative fare with politicians at the Royal Winter Fair in Toronto. In a *Canadian Business* article on Entomo, reportedly "North America's largest cricket farm," journalist Carol Toller marvels at how easily and delightedly nine-year-old Kayla Goldin, daughter of one of the three brothers who operate the business, scarfs down a handful of waxworms. Toller parades out the usual rational arguments in favor of eating insects: on a weight-for-weight basis, they use a lot less land and water than other domestic livestock, and on a per-weight basis, they convert feed inputs into meat more efficiently than, say, cattle.

The title and subtitle of the article get at the primary challenge in scaling up, however. "How a New Wave of Food Entrepreneurs Hope to Persuade Us to Eat Bugs," announces the title. The text beneath reads, "Crickets might just be the miracle food for a hungry 21st century. The only catch? Convincing squeamish shoppers."

Entomo is just a few hours down the road from where I live, near the small city of Peterborough, Ontario, so, in the summer of 2015, before I set off on my entomophagical world travels, I drove over to see for myself what was happening. On my way, I noticed a sign beside the road that read "This land is our land. Back off government." If I didn't know rural Ontario better, I might have started worrying about rural survivalist gangs and whether insects were a menu item for them: Cormac McCarthy meets the Organic Prepper?[87]

I don't know what I was expecting, but there was nothing special that would broadcast this as a cricket farm, the way one might recognize a dairy farm or a feedlot by the sight of it, or a pig farm by the smell of it. The farm building looks like what it is — a converted, warehouse-like chicken barn, about ten thousand square feet, set in a very lightly rolling countryside, a patchwork of boreal forest, corn, and pastures.

The business was started by Darren Goldin along with his two brothers, Ryan and Jarrod; partners Caryn Goldin (culinary manager) and Stacie Goldin (media specialist) are also clearly important members of the team. Darren greeted me as I pulled up to the large open door of the building, a windrow of frass bags along one side of the asphalt parking lot. The former occupant, a now-retired chicken farmer, lives on the neighboring farm and is apparently happy with what these new insect farmers are doing.

Darren Goldin had almost finished his undergrad degree in environmental studies at York University in Toronto when he decided to head west to join a protest against the clearing of old-growth forests in Clayoquot Sound, British Columbia. After that, he and a friend decided to build a cabin in the Kootenay mountains and live off the grid. Finally concluding that he was missing social and family connections, Darren returned to Ontario. He and his partner, Caryn, made percussion instruments and then got into producing insects for pet reptiles and amphibians. Then came the 2013 FAO report on insects as the future of food. That was also the year that the Aspire group from McGill got their million-dollar prize. Darren talked to his brothers. "Hey, we didn't get any prizes," they said, "but we already know how to do this. What's stopping us?"

For Darren and his brothers and their partners, Entomo became a way to translate concerns about environmental degradation into useful action, creating ecologically sustainable alternatives to big-eco-footprint agriculture. I have been on a lot of livestock farms — beef and dairy cattle, chickens, turkeys, ducks. I like the companionship of cows, their low-voiced coughing and grunting, and am happy to sometimes enable them to fulfill their ecological job descriptions, but a cricket farm — the chorus of male crickets singing for a mate — was a whole new experience for me.

Darren walked me through the process. Eggs were grown on clean, slightly moist peat moss. Set on top of this was a cardboard structure of interlocking columns that looked like the kind of thing one would pack wine bottles in to keep them from breaking. The hatchlings would climb upward as soon as they were born, an instinct that seems to be unrelated to light.

From the "nursery," the hatchlings were moved (in their interlocking columns) to blue storage boxes, which were stacked on shelves in a long, narrow room. They were fed a mix of ground corn and soy (grown on the farm), which was spread on the bottom of the blue plastic tubs. For two weeks, as they went through a couple of molts, this was their home: condos for crickets.

Darren then guided me into a much larger room, where the crickets were taken out of their blue boxes, and the cardboard columns were set in rows on either side of long, rubberized water troughs, fed by a drip system. In this room, warmer than the rest of the building (about 30 degrees Celsius), the crickets would stay for four weeks, growing to maturity. I watched them lining up along the thin stream of water, like any other animal, sipping. I felt at home here, amidst the quiet skittering of tiny animals. They were being fed the same feed mix as before. As they matured, the males would begin to chirp out for mates and then (assuming they found a willing female) to breed. Pregnant females would then look for places to deposit their fertilized eggs. Egg-laying places were provided in the form of shallow trays with slopes at either end and a bed of soft peat moss. The eggs would be harvested for use in starting a new generation. If you put the whole tray into a bucket of water, the moss rises to the top and the eggs sink to the bottom. These eggs can then be put into the nursery, where they hatch.

Shortly after mating and laying eggs, crickets die. Just before this happens, between breeding and death, it is time to harvest. The crickets are shaken from their cardboard towers into blue boxes and then dumped through a metal funnel into bags on a weigh-scale — five pounds at a time. Dry ice is added, and the crickets quickly die from a combination of cold and lack of oxygen.

Once the crickets are harvested, Darren said, the water troughs are flushed first with chlorinated water and then with fresh, clean water. The frass is swept up from the floor and kept in sacks. The few crickets that are left run for a "hotel" (one of the cardboard interlocking arrangements) from which they are shaken out into a bag.

Back outside, Darren waved at the bags of frass stacked up next to a fence. The frass, he said, was excellent fertilizer, high in phosphorus and potassium. He has been using it on their own crops and selling to area farmers, and he told me that I could take as much as I wanted for my garden. They didn't yet have a marketing plan, and he wished someone would pick it up as a business venture. They were already so busy just keeping up with the demand for their primary products.

The crickets were processed at a different site, a building in the nearby town of Norwood. Just around the corner from some restaurants and other businesses, there was no smell emanating from the small building, and the neighbors appeared to be just fine with Entomo being there. Inside, beyond a reception room and some offices with computers, was a room with stainless steel ovens like pizza ovens (many drawers, crickets on large cookie sheets). Derek Delahaye, Entomo's processing manager, was washing off batches of crickets in the sink and spreading them on the cookie sheets. If the crickets are to be flavored, that will happen at this point. Otherwise they are toasted and then ground up for "flour." The biggest market for this flour is companies that produce energy bars and protein bars. Derek rattled off a list of at least half a dozen such companies.

Besides Derek, two other people were in and around the kitchen area, a young man and a young woman about high-school student age. For a while they sat at computers, then they moved around the room packing things. As with the other insect businesses that I'd visited in Canada and Europe, I was struck by the fact that this phenomenon seems largely driven by people in their twenties, thirties, and forties. Given that this is the generation inheriting the unsustainable agri-food mess we Boomers have created, it seems appropriate that they are the ones coming up with innovative solutions.

Standing there looking over the cozy kitchen, I popped a few crickets into my mouth: my first *bona fide* cricket snack. Then, as we talked, I tried more, some straight from the ovens and others that were flavored (BBQ, Honey Mustard, Moroccan) before roasting. Once I got past the eyes and legs, and the *idea* of bugs, and popped them into my mouth, I concluded that they tasted, as Darren had warned me, like food. A little nutty. I had expected the legs to stick in my teeth, or trigger my gag reflex, but it all crunched up nicely. Healthy snack food, Darren informed me, crickets were high in protein, omega 3 fatty acids, B vitamins, calcium, and iron. The mealworms, higher in (good) fat, had a slightly richer flavor, like potato chips.

In the summer of 2015, Entomo was producing 4,000 pounds (about 1,800 kilograms) per week of crickets and 1,500 pounds (700 kilograms) of frass, and using about 30 gallons (just over 100 liters) of water per day. I asked Darren if crickets — like pigs, chicken, and cows — had any disease problems. He nodded. Cricket farmers in North America used to raise *Acheta domesticus*, which grow quickly to maturity and have a good feed conversion ratio. In 2009, an epidemic of *Acheta domesticus densovirus* put half of the cricket producers in North America out of business or forced them to select different species. Darren knew of one farm in Alberta with the virus, and the farmer almost dismantled his barn, did a thorough cleaning, and got fresh crickets; in nine weeks, they were all dead again. They now use a different species of cricket, *Gryllodes sigillatus* (the banded cricket), which is resistant to the virus but doesn't grow as large.

Why crickets? The rationale, not only for Entomo but also for other similar companies, is that these are the insects most likely to appeal to North American palates and preferences. If entomophagists wanted to make a big difference, normalizing insects for the North American palate, it wasn't going to happen in $500-a-plate upscale restaurants. As whole animals, crickets and mealworms are like pub food — popcorn, peanuts, chicken wings. They can easily be made into protein powder and used to fortify soups, breads, and energy bars.

In searching for marginal sources of renewal, we can look historically at the origins of our current ways of securing food, and geographically or culturally at places where the modern system has not yet eradicated traditional practices. And then we can look at ways that these different approaches can skittle into the system and change it. These options are not mutually exclusive. If humans have any hope of finding homes on future earth, we will need to continually wrestle with the 'unresolvable' tensions between maintaining some version of an agri-food system and the need for eco-social diversity. The only animal — the only world — without tensions is a dead one.

To come back to Cahill's formulation of the issues, the problematic beast whose margins we wish to define ranges in size and shape from food security for nine billion people to the somewhat more ambiguous monster of modern urban life itself. At a very simplistic level, the margins could be defined by the insects themselves, since insects are at the margins of what people who run the globalizing agri-food system consider edible. There are those who see insects as a way to improve the efficiency of the system as it is now configured, and those who, in the spirit of Michael Pollan's critiques, see the current system itself as a trap. These latter proponents argue that entomophagy offers more revolutionary, transformational possibilities. In this view, eating insects can both bring down the old system and bring in a new way of eating. And because the old agri-food system is so deeply entangled with what we think of as modernity, this new way of eating offers the possibility of profound changes in how we live on this planet.

SHE CAME IN THROUGH
THE CHICKEN WINDOW

*Insects as Feed in*

*Non-Insect-Eating Cultures*

· · · · · · ·

*We may not see them, but they are here*

In 2015, more than forty researchers from eighteen countries published a review titled "Protecting the Environment through Insect Farming as a Means to Produce Protein for Use as Livestock, Poultry, and Aquaculture Feed." They argued that "international fisheries are being over-exploited and current practices are not sustainable, which is evident as current production of fishmeal and fish oil has decreased from 30.2 million tons (live weight) in 1994 to 16.3 tons in 2012. Alternate sources of protein are therefore urgently needed to sustain the aquaculture industry."[88]

Paul Vantomme, one of the people most active on the entomophagy file within FAO, said in an email to me that "the major innovation with insects will be their use as animal feed (aquafeed and chickens mainly) (and with the insects fed on organic waste streams that do NOT compete with human grade foods)." Similarly, Alan Yen, editor-in-chief of the new *Journal of Insects as Food and Feed*, wrote to me that he was observing a shift of emphasis, at least at the insect production end, toward making protein powders that could be used in either animal feeds or products for human consumption.

Enterra Feed in British Columbia, Canada, is considered one of the global leaders in the insects-as-animal-feed business. In May of 2015, since I was already in Vancouver for a public health conference on linking ecological and social determinants of health, I called them up. I was, at that point, naive and ignorant about the business side of entomophagy, but I'd read their website, which sounded promising — but then, what good is a website if it doesn't sound promising?

According to the Enterra website, the company was born when internationally known environmentalist David Suzuki and business entrepreneur Brad Marchant were fishing on the Firth River in British Columbia. Suzuki worried out loud about the unsustainable practices of depleting wild fish stocks to make fishmeal to grow farmed fish.[89] When Marchant asked him what would be better, Suzuki "pointed to the end of the fishing rod, 'How about insects and their larvae?' And thus began the journey of Enterra and the genesis of Renewable Food for Animals and Plants™."

The timing was perfect. In 2014, the City of Vancouver passed a law that required all organic waste to be recycled. It would come into force in 2015. Some businesses cried foul, saying it was not workable, but some of the larger grocery stores, who generate tons of pre-consumer waste — old broccoli, cabbage, ready-made salads, overripe fruits — were thinking more creatively. They could set up their own biofuel systems as some municipalities and farms have done with feces and food waste — or, they could sell their waste to someone who would do it for them. Enterra was ready.

The original Enterra office, located in a low-rise building in Vancouver, was where I met Andrew Vickerson, the Chief Technology Officer. Vickerson is a short, fit thirtysomething with a neatly trimmed beard and moustache and a ready, slightly bemused smile, as if he can't quite believe this is all happening to him, but is comfortable going along with it. He grew up on a hobby farm, studied aquaculture, and then worked as a volunteer in Cambodia looking at raising fish in rice paddies. He figures that this stimulated his interest in insects as feed for fish. I mentioned the insect traps I'd seen in Cambodia, and he raised the problems of by-catch in foraging for edible insects and the possible loss of biodiversity if crude methods were used. He compared it to what had happened to the ocean fisheries.

Enterra started in Vancouver bringing in food waste and processing it in their original facility. Despite its small scale, the neighbors were a little wary. So they moved their farm up the Fraser valley to Langley, where a greenhouse nursery owner invested in their business and let them use his property and old greenhouses. At first, the local government was not sure that insect farming was a "real thing." Officials initially saw the farm as a transfer station for organic food waste, a way to sneak Vancouver garbage out of Vancouver and dump it in the suburbs. In the end, though, they were convinced.

The farm is located just past residential developments of row houses and cheek-by-jowl detached houses. What greeted me as we drove into the paved driveway was a row of large greenhouses and barn-roofed, open-sided storage sheds, belts, conveyors, and pipes. Dump trucks loaded with a sweet-and-sour-smelling wet mess of carrot peels, old fruits, and vegetables pulled in and were weighed full, then dumped their loads. Later they would be weighed empty. Enterra took the "waste," ran it through some mixers to make a slurry, and then used it to feed black soldier fly larvae (*Hermetia illucens*).

I watched the adult flies hatching out in a tall screened-in area. The adult males were doing a kind of spiralling circle dance, first alone and then around the females, who each chose one with whom to to breed. While breeding they were stuck together, like dogs. I saw one copulating couple flying around like a two-ended helicopter. The females then laid eggs into plastic trays that looked like tiny honeycombs. I asked why the flies laid eggs there and not elsewhere. Andrew looked at me and smiled: trade secret. This was a response I would encounter often over the next few months. A lot of angel investors are pouring heaps of money into these ento-start-ups, so it was no surprise that the people involved were reluctant to describe any potentially money-making processes in any detail.

I guessed that Enterra was using a pheromone attractant of some sort. From my quick tour, I understood that fresh fly eggs were put into large flat trays of the kinds that young nursery plants are started in. These trays were stacked, and it took about three or four days for the eggs to hatch. The larvae were fed purchased brewers grains to start. They were shifted to the food waste diet as they went through four molts, at which time they were big enough to pupate and to harvest. Then the cycle started over. The overall process took about a month. The larvae were grown in staggered batches, and each batch was fed daily until they were of harvestable size. A few adult flies flitted around the old greenhouse, lost, having escaped their screened breeding cages, and I saw one satisfied-looking bird perched inside the door. I asked Andrew about escapees, and he said that they contained them as much as possible, but the flies were not an invasive species, besides which the adults didn't feed at all, living only a few days, and the larvae fed on organic matter and detritus.

Andrew told me that 5 kilograms of fly larvae can process 100 tons of feed to produce 6 tons of fertilizer (frass and pupae casts) and 6 tons of larvae for protein supplements. The organic frass was used by local farmers, greenhouse operations, and home gardens, and some experimental data indicated that the frass had insect-repellent or insecticidal properties that kept other pests away from gardens and crops. This would not be surprising since, in evolutionary terms, it would have enabled more larvae to survive.

So Enterra was getting paid to take the substrate (organic waste) and then using it to produce high-quality feeds for chickens and fish. They used only a small amount of land to produce a large amount of protein, and they did not need any extra water; since they could recover water from the fruits and vegetables used as feed, they were actually net water producers. Enterra's website asserts they can recover 4 million gallons of fresh water annually. Again, according to their website, this compares to the 1,400 gallons of water necessary to produce a pound of beef, 500 gallons for a pound of pork, and 400 gallons for a pound of chicken. I asked Andrew about producing insect proteins for human consumption. He laughed. There had been so much paperwork to get their facility registered with the Canadian Food Inspection Agency (CFIA) as a legitimate, licensed producer for animal feeds that he wasn't about to start working on the forms to produce food for people.[90]

Not even taking into consideration the ecological benefits, it sounded like a sweet business deal to me. I looked around the small, rudimentary lab. Most of their more sophisticated lab work, Andrew explained, was done elsewhere. The workers I saw looked to be in their thirties: the future had arrived.

The greenhouses and food-waste trucks at Enterra conformed at least somewhat to my expectations of an ecologically based insect-producing farm, an integration of Suzuki's environmental credentials and Brad Marchant's business acumen. As their website declares, "At Enterra, we're on a mission to secure the future of the world's food supply by solving two major global problems: food waste and nutrient shortage."

If Enterra fit into my general notion of feeds and farming, Ynsect, which bills itself "the insect company," took me by surprise. Their headquarters, located in a secure building in a research park on the outskirts of Paris, represented quite a different narrative for those who see a glowing future for insects in Western societies. Ynsect is a biotech entrepreneur's dream. Their website announces that "this one-of-a-kind technological solution combines industrial-scale insect farming with their transformation into useful molecules for the nutritional and green chemistry markets." Ecologically motivated, perhaps, but this was not a script written by David Suzuki.

I had arranged a meeting with Ynsect's CEO, Antoine Hubert, who has an education in agricultural engineering and experience with managing the wastes from pulp mills, slaughterhouses, and the oil and gas industry. He had become interested in new technologies to reduce and recycle organic wastes, and, from that base, he worked with three others: Alexis Angot, Fabrice Berro, and Jean-Gabriel Levon. In 2011, the four founded Ynsect with the goal of developing what they called an insect bio-refinery. After a pilot study, they expanded their goals and their financial base.

As we walked along the clean white hallways at Ynsect's headquarters, and Hubert talked about their first big investors and their entry into the market — pet food — I recalled how my wife and I used to joke that when we were old, we could eat cat food, which is more carefully monitored for nutrient content than most foods targeted for human consumption. Having previously had some of my research funded by Mars, Inc., makers of Mars™ chocolate bars and pet foods, I am well aware of the kinds of money generated by pet food markets.

Although in August of 2015 they were still at the "promising development" stage, Hubert explained that they were aiming to produce large volumes of reliable insect-based products, efficiently and at a competitive price. The intent was to develop a zero-waste process, using robotics, embedded sensors, and data derived from standardized research protocols. He showed me around the premises, which had been, until recently, part of a human genome lab: high security, proper airflows, locked doors. This was where they would do all their experiments on different genetic species, strains, feeds, and management. The image was of a clean, high-tech, futuristic company, about as "green" as any high-tech agri-food company could be.

Hubert and his colleagues had calculated that, using their state-of-the-art technologies, they could produce 10,000 times more protein per hectare and use 100 times less water annually than if they were using other animals. Their feed inputs would be wastes from the agricultural and food industries. Their processes would be low-impact. Unlike other insect-based feed companies, like Enterra, who used flies, Ynsect had chosen to work with beetles, arguing that beetles had higher protein and lower ash content than flies; furthermore, they had higher chitin content, which could be used to create high-value chemicals and phar-maceuticals. As if heading off a question he knew was coming, he said that they had no interest in GMOs and used standard breeding and genetic selection procedures.

I asked him about end products. The big markets right now, worth billions of dollars, he explained, were for high-quality, standardized feed inputs for pets, fish, and chickens, as well as other insect-based chemicals, and pharmaceuticals. They would later expand into human food ingredients. Working with a Singaporean investment company, Ynsect was expanding its reach globally, developing commercial farm "platforms" in Asia and North America as well as Europe. It was with the promise of these high returns on investment that they had been able to persuade venture capitalists to bring millions of Euros to the table. Still, because of EU regulations, they were restricted to pet foods — not a small market, but not where the long-term ecological and economic gains were going to be.

Listening to his enthusiastic explanation of their vision, I wondered if Ynsect represented the transformative, disruptive, almost revolutionary force Hubert imagined it to be, or if it was something more reformational, a new way to think about inputs to the animal-feed and pharmaceutical industries. I wondered if there was a difference between these two ways of looking at bugs in the system. Maybe they could complement each other, leading to some sort of sustainable, convivial, eco-friendly version of Leon Trotsky's permanent revolution.

Before I left, I asked Hubert who they saw as their main competitors. Enterra Feed was right near the top of the list, along with AgriProtein, a South African company that, like Enterra, was focused on insect proteins for animal feeds. Like the other companies in this hilly landscape of rising and falling start-ups, AgriProtein cites both ecological and economic reasons for their focus, with a slight tilt in emphasis to the economic side. Their website, for instance, explicitly puts their business in the context of soaring prices for soy and fishmeal proteins.

Given their very different approaches and geographic locations, I wondered why Enterra would be a competitor for Ynsect in Europe. EU regulations, Hubert answered. And astute business moves. While Ynsect was working with Singaporean investors, Enterra had come to Europe through the political side door — Switzerland.

In June 2015, Enterra Feed sent out a press release announcing a joint venture with a company based in Switzerland, which is not an EU member. "We are very pleased to be working with Entomeal to expand commercial operations into Europe," said Enterra CEO Brad Marchant. "Entomeal is already well advanced in the use of black soldier fly technology to convert waste food into valuable feed ingredients and their founders have proven track records in business and the feed industry. Our world-leading technology, combined with their local expertise through this joint venture will verify and catalyze Enterra's growth into the region."

This joint venture would be restricted to the small Swiss market, however, until the EU regulations were rewritten. I'll come back to those regulations in a later chapter, as they are likely to strongly influence the shape and direction of the insects-as-food-and-feed movement for the near future.

In October 2015, Hubert announced that Ynsect had developed a feed from yellow mealworms (*Tenebrio molitor*). In clinical trials, the new feed improved FCR such that fish weights increased 30 percent on the same feed inputs. The new feed, they announced, could completely replace fishmeal for juvenile salmonids. If it were introduced in a big way, and the new feed performed as well the trials suggested they would, this could provide some relief for the wild fish populations now being ground up to feed their farmed cousins.

Some might see companies like Ynsect as being more sustaining than disruptive. Are they revolutionizing the way we produce and distribute food, or are they simply taking Cahill's argument to heart and co-opting it, giving it a slight twist so that it serves to strengthen global agriculture as we currently think of it? Is this distinction meaningful as we face an uncertain future in a rapidly changing, perpetually self-disrupting world? Perhaps sustainable food security is best served by a diverse, jostling web of householders in Laos, high-tech business in France, managed mopane forests in southern Africa, intensive palm weevil producers in Uganda, family farms growing crickets for the North American market, and corporate waste processors producing feed for aquaculture.

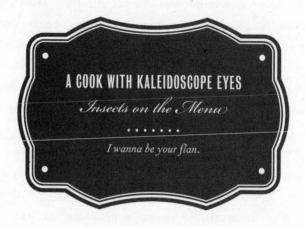

*The Insect Cookbook: Food for a Sustainable Planet*, published in 2014, showcases examples of how recipes from Peru, China, Mexico, and Japan could be adapted to European culinary imaginations. In one of the introductory sections, former Secretary-General of the United Nations Kofi Annan wondered if, a decade from 2014, Thailand would be shipping insects to Europe and elsewhere, the way Argentina and Australia shipped beef. The recipes in the book range from guacamole, chocolate cupcakes, and pizza, to Hopper Kebabs, Bitterbug Bites, and Crickety Kale Salad. In some recipes, the bugs blend in so that finding them becomes a game of Where's Waldo? In other dishes, they are front and center, their multifaceted eyes staring out in uncomprehending blankness. Beetles for every occasion and every taste. One cannot fault the authors for lack of imagination. The endorsements are enthusiastic, the explanations clear, and the recipes attractive. And yet, in the memorable words of Shakespeare, methinks the lady doth protest too much. Can enthusiasm and colorful recipes really carry entomophagy into the mainstream?

In 2012, an incident involving Starbucks suggested that insects' path into non-bug-eating kitchens would require something more than just interesting recipes. That year, the coffee shop chain revealed that the red coloring in its Strawberries & Crème Frappuccino was derived from cochineal insects (*Dactylopius coccus*), causing a frenzied outcry from many of its customers. Some of the objections came from vegans. Some objected for religious reasons, arguing that cochineal insects were not kosher, an argument that creates some interesting dilemmas for scientifically minded Jews. In the late 1920s, Israeli entomologist F.S. Bodenheimer led a field investigation into the nature of the so-called *manna*, which kept the Israelites going in the desert for forty years. His conclusion, that "there is not the slightest doubt that the producer of this manna is an aphid," went quickly from being esoteric research to making global headlines. Aphids are in the same sub-order as cochineal scale insects, which produce the rich red dye that kept the Aztec and Spanish rulers in royal crimson and later caused so much trouble for Starbucks. Not being considered kosher, it raises the question as to whether the Israeli Yahweh was playing a joke, offering up as nourishment an insect that the people were strictly forbidden to eat. Or maybe the Fire that first announced itself to be a nameless "I am that I am" was ambivalent about all those rules and the attitudes they reflected.

Some Frappuccino drinkers objected less for reasons of offended morality, however, than because they preferred their drinks (served in fossil fuel–derived plastic containers) to be free of — like, ewww, gross — things made from bugs.

Whatever their motivation, all of these objections ignored the fact that the ground coffee itself probably contained many more insect parts than the Frappuccino. Most federal agencies that regulate food safety regard insects in food as an aesthetic issue and allow measurable but not visually detectable levels of insect parts in all foods.

When I started working on this book in the spring of 2015, restaurants advertising insects on the menu and start-up companies raising and marketing insects and insect-based products seemed to be popping up everywhere. Entomophagy was a culinary tsunami that had begun in small tide pools throughout the Global South, and a few places in Europe, and was now sweeping the world. Soon, however, even as new ideas, innovative crowd-funding, and prize-winning initiatives appeared, just as many would disappear, or no longer have items available.

In 2008, Meeru Dhalwala, co-owner and chef of celebrated Vancouver Indo-Canadian restaurants Vij's and Rangoli, gave crickets a "soft" entrance onto her menu by grinding them up and making cricket paratha. Three years later she "took the training wheels off" (her words) and advertised cricket-topped pizza. In an interview with the online journal *The Tyee*, she made crickets sound mouth-wateringly tasty, describing them as having "a sort of grassy, earthy, almost nutty, truffle-like flavour."[91]

> *I prefer to roast them, because it keeps them crispy.*
> *Crickets are gentle and have a delicate texture, so*
> *I want to use as little oil as possible. It depends on*

*how you cook it. When cooked whole, they're like
nuts in terms of texture, and need to be cooked as
such. . . .*

*When I cook, I visualize. When you don't
know what to do with it, find a starting point. Stay
anchored to a base flavour. For me, when it comes
to food, my anchor is where my food comes from,
and then from there I try to match it with my personal repertoire of spices. That for me is the most
natural way of cooking. Keep anchored to the base
flavours and then adapt it to your own flavour pref-
erence or palate.*

*Try roasting or sautéing them and test out fla-
vours. I love cumin and cilantro and garlic with my
crickets. Test out what works for you.*

With that sort of PR, how could anyone not love crickets
on the menu? I wrote to Dhalwala asking about her experience
with introducing insects to the menu and expressing my desire
to come to Vancouver and try her fare. She replied that they
had temporarily pulled insect-based items from the menu. As
part of the reason for this action, she cited the kinds of cultural
dilemmas being faced around the world as a corporatized food
culture invades and displaces traditional fare:

*The Indian community has been the most difficult
in terms of introducing crickets — many find it
embarrassing and 'dirty.' I had a much more popular response in Seattle than in Vancouver because
I had a much younger Indian crowd who had been*

*in the US for quite some time. The majority of the Indians working for Amazon who were in the US on temporary two-year work permits were aghast that I would serve crickets at an Indian restaurant, but many were also aghast that I didn't have a cheap $6.99 all you can eat buffet.*

*The media response in both Vancouver and Seattle was great. The hard part is getting customers to order the dish. This is definitely a new food for a younger crowd.*

In follow-up emails and interviews with the media, Dhalwala also underscored the challenges of incorporating environmental and sustainability issues into a restaurant menu. Many customers are looking for a particular curry dish, and are less interested in how it was sourced. Wild-caught crickets could well have pesticide residues, and organic farmed crickets are difficult to find. These issues are not at all unique to Indian restaurants or people from India. I grew up eating the cabbage borscht, rollkuchen, *verenicke*,[92] and sausages that my parents had eaten in the Ukraine; where the ingredients were grown was less important than the "authenticity" of the dish. As a person living in a temperate zone with a long winter, I am constantly running into recipes that claim to be "healthy" and "sustainable" but that call for "fresh, organic" ingredients that would only be available in Mexico or California.

The restaurant story from Vancouver offers a cautionary tale, reminding us that enthusiasm alone won't carry the entomophagical day and that culinary shifts are embedded in complex cultural perspectives and trends that transcend technology transfer.

Where were the inroads being made that might indicate that entomophagy was following sushi's path into global cuisine? Several restaurants have made big entomophagical splashes on the internet windscreens, and as I investigated further, it was obvious that I could spend my life wandering the world sampling bugs cooked up in a variety of tasty ways. Maybe, if I were forty years younger, single, and independently wealthy, I would try that route, but I am none of those. I had read the headlines about Noma, a trend-setting gourmet restaurant based in Denmark: "World's Best Restaurant Noma Brings Insects Back Onto the Plate" (*Forbes*), and "Chef René Redzepi Believes Insects Will Solve the World's Food Problems" (*Eater*). Noma, however, led by what one *New York Times* reviewer called the "godfather of the New Nordic movement," was charging prices that were trend-setting in ways that, were I to sample their cuisine, would require me to remortgage my house and plan my dinners a couple of years in advance.

San Francisco's Don Bugito describes itself as "an edible insect street food project." Their website announces that they "offer rather unusual but tasty creative foods inspired by Mexican pre-Hispanic and contemporary cuisine with locally sourced ingredients."

A restaurant review by Silvia Killingsworth in the August 24, 2015, issue of the *New Yorker* describes another Mexican-derived insect dish, the guacamole at the "Black Ant, in the East Village." According to the report, the house guacamole may include garbanzo beans, fried corn, orange slices, jicama, radishes, and even cheese. "But," says Ms. Killingsworth, "it is always finished with ants. The garnish, to be precise, is *sal de hormiga*, or salt with ground-up *chicatanas* — large, winged leaf-cutter ants, harvested once a year, in the Mexican region of Oaxaca. The ants taste somewhere between nutty and buttery, with a chemical tang, and lend the salt a bit of umami."[93] Killingsworth also remarks on the mixed messaging delivered by the art on display at the Black Ant, featuring, for instance, Salvador Dali's "repulsive ants."

Online, Billy Kwong Restaurant in Sydney, Australia, sounded like an Aussie version of the Black Ant. According to a 2013 interview, chef Kylie Kwong was enthusiastic about integrating varied and delicious insect dishes into the restaurant's eco-friendly Australian-Chinese cuisine. In that interview, which emphasized Kwong's journey from entomophobe to champion of insects as a sustainable and delicious food source, she said that "they're all an integral part of the Billy Kwong menu now."[94]

I visited Billy Kwong two years after that interview, on a warm Tuesday evening in October. The darkly lit place was bustling: young, beautiful, ever warm and friendly waiters, bartenders, and various other workers slipped and danced smoothly past each other, among patrons, into and out of the wine cellar and the bar, past the open kitchen, never colliding — a ballet of black-clad grace, food, and alcohol. The atmosphere was warm and friendly, coziness without claustrophobia, inclusiveness without intrusion. The menu announced, "Where possible we try to use sustainable organic, biodynamic and locally sourced produce."

The Chinese-fusion dishes included beef, prawns, pork, the usual rice and eggs, and some "Australian native greens" I hadn't seen before, such as Warrigal Greens and Saltbush. Insects were nowhere on the menu.

"Oh, I'm sorry," the waitress responded to my query about the bugs. "We don't have any insect dishes right now. Is there something else you would like?"

I was persistent, annoyingly so, and the waitress finally said she would go check with the chef. In the meantime, would I like a drink? After I had a couple of gin and tonics and made some increasingly insistent queries, the waitress returned with what she described as "fantastic news": the chef would make me crispy fried wonton and sweet chili sauce with crickets.

The wonton were indeed a crispy fried pastry, wrapped around prawns, with a light sprinkling of cricket bits on top. I had to take out my iPhone and shine the flashlight on the dish to find them. *Well*, I thought, *they really did kill them when they were little*. The wonton were the perfect combination of crunchy on the outside and soft in the middle (as the polar bears in a Gary Larson cartoon said of an igloo). I ordered a plate of the crispy fried wontons with prawns and no crickets, just for comparison. Really the *only* difference was the sprinkling of cricket bits over the wonton, with no discernible difference in flavor. I asked my waitress if this was how the chef usually prepared insects, and she informed me that they had to be subtle about introducing such new cuisine to Australians, so yes, she only included small amounts.

Kylie Kwong's experience was echoed by that of Meeru Dhalwala in Vancouver. In her *Tyee* interview, she said, "I think in Vancouver, I'll just quietly put it back on the menu when the time comes. I feel like putting too much attention on why people don't like something, it subliminally reinforces the fact that you don't like it.

"I'm going to take a different tactic in Vancouver now. Publicity worked in Seattle, but for Vancouver it didn't have the same effect. We had nothing but positive media attention. People talked about the idea of crickets on a restaurant menu, but Vancouverites all reacted with 'Oh, it's intellectually a great idea but . . . I probably don't want to eat it.' This time, I'll present it like an equal member of my menu and see how things go from there."

Later, the waitress at Billy Kwong brought me a small dish of roasted mealworms — about a teaspoon. They were crunchy, slightly nutty, as they say, not at all greasy — and not worth traveling the world for. A package from Entomo Farms in Ontario would have done just as well.

Head chef Damon Amos of the upscale Public Bar and Restaurant in Brisbane has a different take on bugs in restaurants, sitting somewhere between le Festin Nu's "in your face" and Billy Kwong's "so subtle you won't know it's there." The Public, with its high ceilings, cool atmosphere, and windows on two sides, has an open, relaxed feel to it, less intense and more muted than Billy Kwong, but still susurrating with the conversations of well-off urbanites in their thirties and forties. Here, the insects were clearly on the menu but described with understated and teasing phrases: "Kang Kong, Worms" and "Salmon, Manuka Honey, Black Ants." Kang Kong is also known as water spinach, Chinese spinach, water morning glory, water convolvulus, Chinese watercress, Chinese convolvulus, swamp cabbage, and, to botanists, as *Ipomoea aquatic*. This was a bowl of crispy greens with roasted mealworms. The black ants sprinkled liberally over the cubes of raw salmon, freeze-dried Manuka honey, and chamomile pearls added color, texture, and tang to the sushi-like fusion dish.

In an interview with the *Brisbane Times*, Damon Amos said that he saw his role as a chef to be one of "opening people's eyes to the possibilities of entomophagy and helping them get over their squeamishness."[95] Neal Menzies, a university professor in

Brisbane and entomophagy enthusiast, concurred. In the same article, he speculated that the real value of insects in places like Australia would be as consumers of food waste and replacements for fishmeal as a high-quality protein for livestock.

The world of "new entomophagy" inhabited by Billy Kwong and the Public is imaginatively more complex than the Cold War landscapes some of us hiked around in the 1960s. These restaurants — and the new entomophagists in general — inhabit the virtual construct of the post-9/11 internet. Within and among the worlds of danger, terrorism, plundering neoliberal and state businesses, refugees, wars, and extreme poverty, the many other micro-worlds we inhabit invade and alter our minds and perspectives daily, even hourly, and offer many alternative journeys and landscapes. There are the young and beautiful, travelers and activists, neo-urban farmers raising chickens and bees, reinvigorated and redefined indigenous nations.

It's not clear to me yet where the new entomophagists fit into this inchoate mix of evolving global cultures. I suspect that, coming from the margins of all these cultures, they will find many homes, some where crickets are a garnish, where ants or stink bugs add a bit of *je ne sais quois*; some where they are the main protein or energy source; and others where the insects are primarily protein sources for other animals. I am not surprised, and indeed it is entirely appropriate, that some of the biggest success stories on the restaurant end are in those places where, even in the face of modernity, insects have never quite left the kitchen.

For instance, in March 2002, former disc jockey Pailin Thanomkait, thirty-two, and her partner, ex–shrimp farmer Satapol Polprapas, twenty-nine, launched an insect-serving fast-food chain in Thailand. Catering to adventurous Westerners and relatively well-off urbanites, Insects Inter can, and does, charge higher prices than the street vendors because they source their insects from known farms and guarantee them to be pesticide-free. In their public relations statements, they emphasize quality control and a special sesame oil. Within a few years of its founding, the chain was rapidly expanding the number of franchises across Thailand and talked about expanding and moving into China and Korea.

After my bug-nibbling experience at Billy Kwong, I stood for a while at my window, looking out over the city, Sydney's skyline silhouetted against its own glow, the dark, graceful arc of the bridge, the white, accordioned seashell of the opera house. I was — and am — puzzled by the high profile that insect cuisine carries internationally and virtually, and the low, near invisible profile it carries locally, in the daily lives of most people. This has to do with the ways in which social media have often reinforced fragmented understandings of the world; there are implicit — and false — assumptions in each isolated group that others are working from the same visions and knowledge, that there is "One World" and "One Health." All of us humans may see a shimmering mirage of a refuge or castle or Eden on the hill, but our assumptions about how to get there, and what we will do on arrival, are not the same.

While insects as animal feed offer opportunities to make a new green economy economically viable, sprinkling insects into the menu may provide a more open avenue for cultural change. Being a both/and type of person, I see important eco-culinary niches for both approaches, each making a unique contribution to creating a more complex global narrative about people, bugs, evolution, ecology, and sustainable health.

## PART VI. REVOLUTION 1

We love animals as pets, and, if we just can't overcome those deep desires to get our teeth into something animal, we eat them. But this plunges us into quandaries of ethics and rights, policies and regulations, food scarcity and waste overabundance. We already have trouble imagining what it means to interact ethically with dogs, cows, and chickens, but can we talk about cricket welfare? How do we even start a discussion that includes love and bugs without sounding silly? Can this story possibly have a happy ending? Let us see where we might be going, and what a "Day in the Life" might look like if bugs are on the menu.

## IT'S SO HARD (LOVING YOU)
### *Ethics and Insects*
. . . . . . .
*Boy, you're gonna carry that weight.*

Most of those who have explored or extolled the promises of entomophagy — myself included — are neither philosophers nor ethicists. We have backgrounds in agriculture, food security, health promotion, and natural sciences. It is no surprise therefore that ethical issues have received scant attention by entomophagists. The challenges identified by most researchers have been related to public relations, technology, and regulations.

The 2015 article by Matan Shelomi to which I referred earlier typifies the most common way of framing the issues. Titled "Why We Still Don't Eat Insects: Assessing Entomophagy Promotion through a Diffusion of Innovations Framework," Shelomi's article returns to the nagging question of why, after all these years, and all these arguments, we (by which he means

Europeans and their descendants) *still* don't eat insects. In the end, he concludes that scientists who are "interested in developing entomophagy should focus on rearing and packaging insects rather than worry over how to convince others to eat them. Create a safe and steady supply," he cavalierly asserts, "and demand will take care of itself." He also predicts — based on theories of innovation diffusion — that once the bigger farmers with their economies of scale and mass-rearing technologies enter the market, all the smaller farmers will go out of business. Technologies that enable the new, commercial producers to scale up, producing a steady stream of edible bug products, will represent disruptive innovations, transforming the agri-food industry. Food safety and environmental issues, he predicts, will be regulated and managed in pretty much the same way as in other livestock enterprises. For those who wish to see other non-Western cultures keep or increase their traditional insect-eating practices, he suggests that, given "the predominance of acculturation towards Western lifestyles among economically marginal populations, we may see other cultures emulate the new, fashionable trend of entomophagy and add insects to their diet . . . exactly what we wanted in the first place."

If one takes authors such as Shelomi (and many of the participants in the FAO conferences) at face value, entomophagy is seen to be an issue that is best — and sometimes *only* — considered in terms of biology, economics, and nutrition. In this view, the appropriate responses to the challenges of globalizing entomophagy are the development of new technologies and public relations programs grounded in natural science research and corporate models. Gender and power imbalances and economic inequity are externalized, to be dealt with as necessary by politicians as they are for other forms of agriculture. Basically, so the thinking goes, some combination of mass insect production systems with better advertising campaigns will get "us" where we want to go. In this context, it is not clear to me who that "us" refers to, and whether this is indeed what "we" want.

Lest we think nontechnical questions of ethics and animal welfare are of mere "academic" interest when we are talking about insects, consider the case of Florence, Italy, where crowds have traditionally gathered every spring along the banks of the Arno River to celebrate. To the church, this is a celebration of Ascension Day, when Jesus is said to have risen into the skies. To many other people, this is more simply Festa del Grillo, the "Cricket Festival." Some authors have argued that the intent of the festival, historically, when populations were more closely attuned to the lives of farmers, was to reduce the numbers of cricket pests. Others have argued that the festival has its roots in ancient pagan celebrations of spring. Whatever its origins, people used to bring their own crickets, or buy them — chirping in their colorful wooden, cane, or wire cages — from vendors. In 1999, however, the Commune of Florence passed a "protection of animals" law that effectively made selling live crickets illegal. Now you can buy little electronic toys that make cricket-like sounds. Is this progress, or just another step in our alienation from our biological selves? What will happen when such laws are applied to cricket farmers?

If one takes a less naive, more realistically nuanced, attentive, systemic, and ethically grounded view than that expressed in Shelomi's article, entomophagy is not just a "why not eat them?" issue to be solved by some technological sleight of hand. In this more complex understanding, the ethical challenges facing entomophagy are embedded in social relationships (the health and well-being of people, including producers as well as consumers) and biological webs (the welfare of the insects as animals and the health of our shared ecosystems). Rather than trying to clean up the social and ecological messes left behind by visionary technologists, we might consider how to describe the "world we want," and then design appropriate technologies to take us there. How can we reframe the entomophagy debates to better enable us to talk about some of the nontechnical issues?

One of the first questions that came to my mind in approaching this subject was a confusion I had about morality and ethics, which seemed to be used interchangeably in many public debates. I asked Karen Houle, a philosophy professor at the University of Guelph, about how she would differentiate between these two terms. "Morality," she explained to me, "has to do with what we think of as rules for right living. Traditionally, these rules have been framed in religious terms. Thou shalt not. Thou shalt. So a moral person is a person who A) has a set of rules to live by and B) lives by them."

In some cases, such rules are absolute. My father, for instance, held the view that killing was wrong, period. This translated into a strong position against all war, no matter how just the cause, as well as opposition to capital punishment. His position was that it was better to die at the hands of an evil person than to kill them. An absolute moral code can be comforting because you always know, at least theoretically, exactly what to do in any situation. But in many situations — World War II and abortion being two examples for my father's position — this kind of absolutist moral stance generates anguishing dilemmas. For eating insects, with whom we have such an entangled and conflicted history, an absolutist stance creates as many problems as it purports to solve. Albert Schweitzer's philosophy of "Reverence for Life," for instance, led to an insistence on protecting, nurturing, and saving every insect, including those crawling around in the bottom of holes dug for fence posts.[96] Was this *only* a revulsion against killing, or was it more than that? A revolt against the natural world, the ecology we inhabit? Is not killing insects that spread disease in fact a choice to allow children to die from malaria?

I asked Houle about ethics. "Ethics," she said, "has to do with grappling with a situation entirely too large to comprehend, let alone make rules about, let alone figure out which rules to apply. . . . An ethical person senses that there is something in me, in the world, in life, that demands my care and attention and all my intelligence. And I could fuck this all up totally. I have to try, though. I find myself having to try. It is grounded in a sort of built-in primordial capacity for caring, and a capacity for joy and pain and knowing the difference inside oneself, every step of the way. So ethical life consists of finding ways to carefully put one foot in front of the other, without a map, and, on the way, not stepping on anything you find that you care about. All you know is you have a foot, you have feelings, others might too, at the very least they have lives, and all these lives fit together to make our one shared reality. And life is short. Your capacity to live well (i.e. feel joy rather than puny or sad) and other things being able to go on, in their way, with their beauty and strange ways of being what they are is somehow augmented by caring and trying rather than not. The opposite of an ethical person would be an apathetic person. A-pathos. Someone who is already dead inside their world."

Regulations reflect societal aspirations and ideals. They are to those ideals what morality is to ethics. The best technology links the ideals to the regulations. This brings us to questions of whose ideals we are considering, and to my suggestion that our technology should be informed by the world we want. But who are "we"? The moral codes we apply to animals are culturally rooted. North Americans eat cows, but in India they've lynched people for doing the same. In the 1990s, when I worked in Nepal, I was told it was morally acceptable to eat buffalo, but not cattle. I recall talking with a very sincere and morally conscientious colleague about the way dogs and cats were caged in Chinese food markets. He seemed surprised that this was even an issue. The people were hungry. The dogs were raised to be eaten. What was the problem?

In an academic paper titled "Jurassic Pork: What Could a Jewish Time Traveler Eat?" the authors explore what they call "kosher paleontology." They found that the answer to their question depended not just on which religious scholars and translators one consulted, but also on more recent understandings of Linnaean classification. It appeared that locusts and crickets would probably be okay. There was no consideration of the possibility of suffering, nor did they mention Bodenheimer's research, which concluded that the biblical manna was a scale insect.

The "kosher" issue is just one example of the general problem with moral rules derived from religious scriptures, which are themselves a special case of ethics versus morality. Religious and ideological experts and authorities cannot agree on what their founding documents or scriptures mean, or what their authors intended, and therefore what the rules should be. As soon as people try to translate apparently simple ethical stances —biophilia, respect for life, or promoting a common good — into a complex, dynamic world, the public debates degenerate into contradictory rules and regulations. The questions raised cannot be resolved by gathering more evidence. We would be fantasizing if we thought that such conflicts could be avoided as entomophagy enters the mainstreams of twenty-first-century urban societies. The outcry against the use of red dye from cochineal insects to color Starbucks's Strawberries & Crème Frappuccino was driven as much by the concerns of vegan absolutists as by those of religious purists.

So, rather than quickly getting mired in the many sets of culturally, politically, and religiously defined moral codes, I think it is helpful to stand back and grapple with the ethical considerations from which these rules have emerged. Entomophagy is an issue that can — if managed thoughtfully — disrupt economic and gender relationships, politics, policies, and regulation. Beyond that, it can challenge our relationships with the biological world, in which we are but one entangled organism among millions and millions, and raise questions of who we are as humans — and who we want to become. In this context, can we agree on some broad ethical considerations? If so, then we can begin to devise an approach to entomophagy that is ecologically and culturally appropriate and can enable us to accommodate different regulatory approaches (a.k.a. moral codes).

Some have proposed that we describe this fog-bound ethical landscape in terms of what some of us would call love. Celebrated entomologist E.O. Wilson speaks of biophilia, the instinctive love of life. The word *love* bears its own burdens, of course. As Michael Ignatieff notes in *The Needs of Strangers* (which insects, to most of us, most assuredly are), "Many of the things we need most deeply in life — love chief among them — do not necessarily bring us happiness. If we need them, it is to go to the depth of our being, to learn as much of ourselves as we can stand, to be reconciled to what we find in ourselves and in those around us."[97]

At least one of the "inspired by real life" fictions by veterinarian Alf White, better known as James Herriot, suggested that the author might love animals, especially dogs. Many North Americans speak without embarrassment about being dog lovers, horse lovers, and cat lovers. But can we talk about loving insects — Black flies? Mosquitoes? Bed bugs? — without drifting into vague similes and metaphors? Can we, without falling into the flake-filled pit of an imagined multiverse, create a language that incorporates and transcends the ecological tongues of pheromones, scent, color, sound, taste, and magnetism that whisper around us and connect us to all other living beings?

The notion of a love that does not necessarily bring happiness is in keeping with the approach to studying and managing ecosystems advocated by Henry Regier, member of the Order of Canada, Professor Emeritus and former Director of the Institute of Environmental Studies at the University of Toronto. In considering how we might frame initiatives around the Great Lakes Basin, Regier argues that love encompasses much of what is important for integrated assessment and management. Love, as he defines it in this context, is a complex phenomenon that includes and transcends other systemic approaches to natural phenomena, such as energetics, economics, and ecology.[98]

Jeffrey Lockwood and Harvey Lemelin have proposed that we embrace our conflicted sense of insects as both awful and awesome, a kind of live-and-let-live attitude that Lockwood calls entomapatheia. When it comes to insects and ecosystems in general, maybe entomapatheia is what Regier's love, and Wilson's biophilia, *feel* like. Wilson and Regier have given us some examples of how their views might work out in practice, from managing the Great Lakes Basin to conserving insect habitats.

For me, the word that approximates how I feel about the natural world, and insects in particular, is *care*, whose proto-Germanic etymological roots evoke ideas of lament, grief, and sorrow as well as a non-Hallmark sort of love. I care about insects as I care about all animals, even when they annoy me. They make possible the world that is cohabitable by different species, even as I lament my necessary ethical entanglements with them.

Caring about nature sounds like a good place to start, but in applying these notions of biophilia to the concept of insects as food in ecosystems, matters quickly get complicated. In other, non-entomophagy-related human–animal relationships, we have spent more than a few centuries arguing about animal conscious-ness, emotions, welfare, and rights, and whether, more generally, humans have any obligations toward animals — and, if so, how far those obligations might extend. The progression of our thinking in these areas has had an important influence on how we organize and regulate everything from livestock farming and research to zoos and wildlife conservation parks. Insects are, by scientific clas-sification, animals. Do the insights and laws developed to manage our interactions with, say, cats, dogs, cows, and pigs also apply to edible insects? If so, why? And how?

If our experience with non-arthropod animals is any guide, then the regulation rubber hits the road when we talk about suffering. If animals are capable of suffering, we say, then we have some obligations to them, a basis for passing laws and promulgating regulations. But why would that be? Why does infliction of suffering lead inexorably to obligation and regulation? Unlike some other veterinarians, I have met animals — and most certainly insects — that I could only love in the non–greeting card sense suggested by Wilson, Ignatieff, and Regier, which popular culture might not recognize as love at all.

Which brings me back to *care*. Even when I don't like or love an animal, I still care about it. At bottom, I don't want to cause suffering in other animals or people because I care about them. I'll come back in the last chapter to *why* we might want to care, but for now, I will assume that we do and move on to what the implications of this caring are.

In my research on entomophagy, I have not come across many people talking explicitly about ethics, or even its more familiar moral cousin, animal welfare. Jeffrey Lockwood, a trained entomologist and now Professor of Natural Sciences and Humanities at the University of Wyoming, whom I mentioned earlier with regard to locust plagues, is an exception. In a couple of articles in 1987 and 1988, Lockwood argued that insects deserve our moral consideration because there is sufficient evidence that they — at least the social insects — are capable of suffering.

Lockwood says that "considerable empirical evidence supports the assertion that insects feel pain and are conscious or aware of their sensations. In so far as their pain matters to them, they have an interest in not being pained and their lives are worsened by pain. Furthermore, insects as conscious beings have future (even if immediate) plans with regard to their own lives, and the death of insects frustrates these plans. In that sentience appears to be an ethically sound, scientifically viable basis for granting moral status, and in consideration of previous arguments which establish a reasonable expectation of self-awareness, planning, and pain in insects, I propose the following, minimum ethic: We ought to refrain from actions which may be reasonably expected to kill or cause nontrivial pain in insects when avoiding these actions has no, or only trivial, costs to our own welfare."[99]

I was skeptical of Lockwood's claim that insects can suffer. Nevertheless, scientific reports in the past few years (since 2013) have documented the deep similarities between vertebrate and insect "brains." One 2016 report in the *Proceedings of the National Academy of Sciences* is titled "What Insects Can Tell Us about the Origins of Consciousness." The authors assert that "invertebrates have long been overlooked in the study of consciousness," and that the "time has come to take them seriously as a scientific and philosophical model for the evolution of subjective experience."[100] The Green Brain Project, funded by the Engineering and Physical Sciences Research Council in the UK, announces that their work "combines computational neuroscience modelling, learning and decision theory, modern parallel computing methods, and robotics with data from state-of-the-art neurobiological experiments on cognition in the honeybee *Apis mellifera*."[101]

If we can use insects to study the evolution of human consciousness and to build computational models, robots, and drones, then I suppose the notion of insect suffering is not so far-fetched after all. Still, where does this lead us? Is Lockwood on the right track?

I asked Houle about Lockwood's ideas and, to my surprise, she was not impressed.

> *Maybe this is the right "moral recipe," but nobody can actually cook with it. Here's why. To get to*

*the conclusion of the experience of suffering taking place, and not just conclude that what was seen (or measured with thingamabobs) was flinching, that is, mere instinct (sensation awareness) is a gigantic leap, both conceptually and empirically. If you try to make that leap you get nailed for being soft-headed (bad scientist or projecting, or both). Or maybe you do manage to rig the objective cortisone flinch-measuring thingamabob and it says "suffering is happening in cricket at time T5." Then, in one or two moves along the graph you can find yourself proclaiming: it passes once their wings are pulled off. The problem? You need to posit a higher-order subjectivity in the being to get insects "onto" the same empirical-moral playing field as human consciousness and human suffering. That means we need to posit (or better, to prove) that insects have a sense of self and a sense of being invested in that self in a certain quality of life and over time. We are already having trouble designing and carrying out experiments to prove, without a doubt, that this is the case with orangutans and gorillas. The empirical prerogative allows people to appear most rational if they continue to doubt the experimental evidence as proving that. So: If you don't have this level of conviction and evidence, i.e. proof about the invisible continuous-over-a-life inside thing called self-consciousness, then a rock falling on you just hurts as it is smashing you. But you don't (apparently) suffer because you don't know that you are the self that didn't want that to happen in the first place, or reflect upon the trauma after. Just bug squash.*

*So much research, with insects, dolphins, cats, mice, and cows is devoted to trying to figure out what quality of "awareness of sensations" they have on the inside, so to speak. We can only read the symptoms (cortisone levels rise in blood tests, facial grimaces, screams). Even then, so many people say: Oh, we can't be sure those are the effects of a cause called pain. But the piece about establishing second-order personhood in order to then say that they suffer in order to then say what Lockwood says has so many black boxes along the way.*

Houle clarified that she respected what scientists like Lockwood were trying to do — taking the empirical, scientific criteria we use to give cute kitties moral standing and asking, can this reasoning be applied to insects? But she was skeptical about empirical science being able to close any of the gaps in that argument and arrive at any firm conclusions. "Think about it," she said to me. "I don't even know with certainty what you are feeling. It's all smart, bodily based, experientially informed statistically sound guesswork with a giant gesture of benefit-of-the-doubt thrown in. That last piece — benefit of the doubt (a.k.a leap of faith!) — is where things are going to slide."

Beyond consideration for the suffering of the animals being managed and killed, there are those of us who, at least sometimes, feel complicit in whatever suffering might be inflicted by rearing and killing animals. We agonize over it. Does the killer suffer? Can this be legitimately called "suffering"? Any conscientious meat-eater, livestock owner, or slaughterhouse worker knows that, as Houle said to me, "the suffering isn't ever 'contained' to the object or target but flows and leaks through all beings who inhabit that space. Life is porous. Life leaks. The human subject lives a life that slides in all directions. A feminist care perspective would say that we are all in this together, and we all can and should care about each other's well-being; lead with our hearts, not with our heads, and that way, feeling (compassion) will be available to us and even the killers will be seen for what they are: people among people."

I thought again about "trivial implications for human welfare." Eating insects because doing so will feed millions of poor people, thus alleviating their suffering, and save the world from catastrophe, hardly seems trivial. In fact, these seemed like big, nontrivial loopholes, reasons to inflict pain, similar to the notion of "just wars" among many people who generally consider themselves peace-loving. When it comes to the crunch, will not human hunger always trump the welfare of crickets? Having passed through this loophole, we may not seem to be any further ahead. And yet, did not an Albert Schweitzer–like absolutism, without any loopholes, lead to its own dilemmas and, some would say, absurdities?

Having acknowledged the unresolved, and probably unresolvable, ethical quandaries faced by our relationships with insects, entomophagists still need to — and *want* to — make good decisions about managing and killing these small, six-legged animals. In some ways, once we accept certain animals as food, the question of how to kill them, however emotionally charged it may be, appears to be simpler. Humane slaughter — an odd turn of phrase for those not in the meat business — of livestock such as pigs and cattle has long generated controversy. Temple Grandin, the celebrated autistic animal scientist, asserts that she can see the world as an animal such as a cow might. Her work, and that of her colleagues, has led to a wide range of improvements that minimize suffering in farm animals during transport and slaughter.

Temple Grandin's approach fits with how many insect farmers treat the humane killing of insects. The question becomes one of how to minimize pain and, insofar as we can tell, suffering. Some would say boiling is the quickest and least painful way for crickets to go. But how do you get a mass of crickets into the pot? Some farmers use dry ice, but others say that insects have a high tolerance for low oxygen. Or freezing. But some insects survive freezing. Flash freezing maybe? Many — though not all — of the insect producers that I spoke with cared about their animals and did not want to inflict pain and suffering on them. Given the uncertainty of scientific knowledge about insect pain and suffering, they would hedge their bets and use a combination of dry ice (which probably kills the insects, and at the very least knocks them out) followed by cooking or roasting (which ensures they will be killed).

In the context of killing animals humanely, some non-vegans argue that hunting deer is less ethically problematic than slaughtering cattle; hunted deer die suddenly, in the midst of life. They have one really bad day in an otherwise good life. Would we not all wish to die that way? For the entomophagists, this argument would favor foraging over farming. If you catch insects efficiently and dispatch them quickly in their home habitat, then, so the argument goes, they "die happy."

However, the nature of the question changes if consumer

demands for such "ethically captured" insects rise, and millions of people go tramping into forests and meadows and swamps to meet those demands. Insect-foraging in an overcrowded, bug-eating world may then result in the death and destruction of millions of other individuals and species as their habitats are destroyed and/or inedible or "undesirable" insects are caught in the nets. In fishing, this is euphemistically called "by-catch," and in war, "collateral damage." We might feel better about ourselves, but in the process of resolving a local, personal problem, we will have left a much larger negative ethical footprint.

We cringe when we see a cow or a dog or a horse suffering. This strongly suggests that something is ethically problematic. When the dog is shot in the movie *Old Yeller* or, more recently, the New Zealand adventure film *Hunt for the Wilderpeople*, we are upset, even if we understand at one level that this killing is stopping the animal's suffering and is therefore the "right" thing to do. It is a quandary many veterinarians have faced when putting injured animals out of their misery. They cringe and yet go ahead with the killing because they are intent on minimizing pain and suffering.

For most people, the cringe factor doesn't seem to arise when we think about killing insects, especially if the killing is done quickly and out of sight. We lack the imagination to empathize with insects. They are so very much *other*. How can I imagine what it is like to be a cricket if I cannot even imagine what it is like to be another person? The cringe factor arises when they arrive on the plate. However, in the entomophagy literature, this cringing at the insect in the salad has been interpreted not as an ethical issue, but as one of European cultural bias, to be overcome with better advertising.

But is it something deeper? A disturbing instinct that suggests something is wrong? A recognition of the unsettling dilemmas posed by having been born into a world that seems to demand a life for a life? It is, I suggest, always worth asking the questions.

Another criterion for ethical consideration, beyond suffering and cringing, has to do with how we value insects' lives. Much of what is implicit in how entomophagists value insects has to do with the ecological services that bugs provide, their usefulness as food, and the disease-transmission threats they represent. But we value many things in life — friends, family, art, music — without any reference to such quantifiable outcomes. In many cases, we value animals that are rare, or in some sense beautiful — the so-called charismatic megafauna like lions and pandas. Insofar as insects are beautiful, they evoke ethical sensibilities in us. But what constitutes beauty in insects?

In some cases, people see beauty because of religious or spiritual connections. Albrecht Dürer included the stag beetle in his 1504 *Adoration of the Magi* because of its association with Christ. In other cases, economic value, play value, and beauty are entangled. In Japan, over a million stag beetles from 700 species are imported annually as pets and display animals. Some have fetched prices as high as US$5,000. In 2004, the twentieth Japan Rhinoceros Beetle Sumo Championships attracted over 300 contestants. In 2001, the number of rhinoceros and stag beetles alone imported into Japan was recorded at 680,000, from twenty-five countries. Some entomologists have even argued that the best place to look for biodiversity of stag beetles is in Japanese pet shops.

In other cases, there is a kind of aesthetic sense of beauty with no commercial or religious overtones. Galleries online include

"The Unexpected Beauty of Bugs,"[102] for instance, and "Beautiful Bugs of Belize."[103] Even some scientists express a kind of awe at the beauty of their subjects. Hölldobler and Wilson's 2009 book *The Superorganism* is subtitled "The Beauty, Elegance, and Strangeness of Insect Societies." Bees are particularly prone to being thought of as beautiful for reasons unrelated to religion. In many parts of the world, people eat the adults and brood of bees and their relatives. In Europe and North America, entomophagists have pretty much avoided asking people to eat bees; it would be like promoting beef in New Delhi or putting dogs and cats on the plate in Toronto. We *like* bees. They are bodhisattvas, democrats, and essential pollinators. *The Insect Cookbook*, published in the Netherlands, lists bees as one of the insects that *could* be eaten, but in the actual recipes, they only appear as marzipan decorations.

So, as for ethical considerations in dealing with insects, whether foraged or farmed, we have their possible capacity for suffering. We have wildly different notions of beauty and value. We have the complicated idea of care. And, once we raise our ethical gaze from the grasshopper in our palm to populations, species, and the dynamic webs of multispecies communities and landscapes, we have what we might call "contracts" or "pacts."

According to Houle,

> *we have made pacts with many animals over the centuries, implicitly (we have adapted to each other) and explicitly: You are my 4H pet, please come home and I will look after you. You can't just break promises and eat your friends because you are*

*hungry. . . . To be in the world, to come into this world, we discover that we are a party to various on-going agreements. Gentlemen agreements (who parks first in a shared drive); formal agreements (I pay taxes, the government gives me back any I overpay); informal agreements (I pick up my litter, you pick up yours; don't come empty-handed to a potluck); natural agreements (don't shit in the well or shit over on the side of the property. My dog knows that one); legal promises (my parents had promised each other till death do us part, before I was born: I was party to that promise).*

*Sometimes I think ethics is nothing more than discerning those webs of pacts we are implicated in, and doing whatever we think is appropriate to recognizing them (even if it means outright refusal to). The bees. The bees do their thing. They keep doing their thing. We benefit in a zillion ways from that. Some of us (beekeepers) directly intervene to enable that. Others (supergiant bee corporations) have the almond companies more in mind, but realize that profits depend on bees. Others (pesticide and herbicide companies) do not have the bees in their sights at all when they are spraying cash crops. The bees drop. So what if we did characterize our relation to them as a pact, a pact of mutual acknowledgment and non-interference? Minimum. Maybe even support (putting out water; building shelter). If every one of us saw the bees as creatures with whom we*

*have a bilateral trade agreement (and our existence,*
*theirs and ours, depends on it being honoured) then*
*maybe we could get somewhere!*[104]

A 2015 research project reported that every year ants cleaned up great heaps of organic street waste — the equivalent of 60,000 hot dogs — from the streets of New York. Many of these ants are considered invasive species from Europe. So we have a pact: you can live here if you clean up the garbage. Does this come down to trade agreements?

Houle laughed when I suggested this.

*Please, no. Not so quick with the Market Model!*
*We also need them for our psychic wellness.*
*Otherwise we would be completely off our rockers.*
*Animals, insofar as they are Other to us, and yet*
*nearby, close, neighbours, co-creatures, can teach*
*us how to be better humans . . . That maybe won't*
*work for snakes, hornets, bed bugs, midges. So*
*that tells us that a mirror phase — that reciprocity*
*(formal or informal) — is available, but only with*
*some other creatures. And does it go both ways? I*
*always wonder that. Does it have to, for this kind of*
*ethics? Yes and again no. There are promises made*
*on behalf of those who seem to not be able to make,*
*or understand, promises. . . . Another consideration*
*in this idea of a pact is that all animals are more*
*vulnerable to us than we are to them. Maybe this*
*fact itself (a power imbalance) dictates that it's not*

*right to then exploit this unlucky station of the vul-*
*nerable by inventing recipes with them in it. (They*
*can't do the same of us.)*

*A very general idea in all sorts of zones of ethics is*
*that exploitation of vulnerability is a wrong. You are*
*a soldier. You are guarding a prisoner. That prisoner*
*is 100% vulnerable to you. If you do anything to him*
*because of that (rape him, mock him, hit him) there*
*is a wrong attaching to your being. Same with chil-*
*dren. Same with insects in the hand of a child. Same*
*with earthworms in the marsh that is being drained*
*for housing. Total unilateral impact in a particular*
*situation. The wrong is to simply act upon the other*
*for one's own ends, however trivial.*

And what if the ends — sustainable food supplies, the allevi-
ation of hunger — are not trivial at all?

If we step back from looking at the suffering of an individual
animal and consider the ecosystem of which it is but a small
member, we are faced with a different set of challenges. One of
the most forceful claims being made in favor of eating insects
is that they have a lighter ecological footprint than traditional
livestock, require fewer resources to raise, and produce fewer
greenhouse gases. But, in insect agriculture, it depends on what
you feed the insects. It's one thing if you are feeding them
waste products, like Enterra Feeds does in British Columbia. It
becomes more problematic if you are formulating rations that
mimic those of chickens.

That old adage "waste not want not" runs through all the entomophagy literature, the sustainable development literature, and the wisdom-of-those-who-lived-through-the-depression literature. Efficiency has become one of our societal ideals. Part of me is attracted to that. It has been my argument for why pot-bellied pigs are an ideal food in Bali. They are scavengers, and if you kill one, it is generally just enough for a family meal. If you make them pets in North America, which in my view is condescending and disrespectful of the pigs, the argument falls apart. Still, I worry about not-wasting as a criterion for ethical behavior. It tends to get reframed as tweaking what I see as an already problematic agricultural system: Can we make chickens more efficient at converting feed inputs? Yes, if we feed them low-level antibiotics. Can we grow cattle more efficiently and reduce waste at the slaughterhouse? Yes, if we feed the offal and other bits of the butchered cows that people don't want back to calves as protein supplements. A single-minded aspiration to be efficient and avoid waste too easily degenerates into a view that places short-term profits above everything else, including unfortunate "side effects" such as mad cow disease and widespread antimicrobial resistance.

*What the threshold is for "waste" or "overconsumption" will change with each situation, each lake, each species, each demographic. Philosopher John Locke noted that nature and the seasons set natural limits on what was available, how much work would be done (input and outputs), and then how much each family could take. Ideally there was a balance in there: input, output, get to the next season, keep enough of the good soil, don't eat the seeds, etc. It has a beautiful toward-the-future feel to it. There is a story about some folks in Siberia refusing to eat the seed stores even when starving. It would be like eating the future. But Locke also noted that gold put a total wrench into that system because you, personally, could harvest ten times more than you needed or could eat in a season, but sell it, so technically you were off the hook for 'wasting.' If the person you sold it to let it rot, that wasn't your problem anymore. Capital has put so much pressure on the production end: fish farming, greenhouses producing and harvesting eggplants out of season, overtaxing the soil . . . that those natural limits are hardly part of our consciousness anymore.*

Thus, while using insects in food and feed as a way to reduce waste in the agri-food system is an admirable goal, the use of bugs in the system-as-is must be constrained by a much broader and deeper understanding of the eco-social context within which this waste is occurring in the first place, and the ethical questions raised by that context.

The late systems design engineer and ecologist James Kay argued that the complexity of the world we inhabit always leads to uncertainty and trade-offs. The problem here is that we're faced with a situation in which we are not sure what the trade-offs are. No matter what we eat or do, we exact costs. By just existing, we are responsible for the deaths of many insects, bacteria, animals, and plants. Even if we go with foraging and "eating wild," there's the problem of overforaging sometimes being more destructive than farming. Even if we don't eat animals directly, we live in their spaces, and we eat foods that they could be eating. Of course, we also make possible other lives, in our intestines, in the habitats we create, in the decay of our bodies when we die.

Insect farmers such as those at Entomo are faced with trade-offs when it comes to water and energy use, disease control, feed conversion ratios, keeping crickets comfortable and "happy," and keeping their own human families fed and housed. Matthew Waltner-Toews, of Unspun Honey in Australia, characterizes his beekeeping style as apicentric — that is, putting the interests of the bees first — as distinct from what he sees as the more common anthropocentric practices. Even after having made such a choice, however, he is faced with trade-offs. He has chosen to use Warré hives, but he has also considered some of the newest innovations for apiarists, such as the much-hyped Flow™ Hive, and hives made from expanded polystyrene. In each case, there are trade-offs between convenience for the beekeeper (ease of extraction, pre-set plastic cells in the comb), comfort for the bees (insulation value, freedom to create their own cells from wax), and long-term environmental impacts (whether hives can be recycled, for instance) that might impact the world his children are growing up in.

This, then, is what I carried away from my exploration of ethics and entomophagy. We will need to consider suffering, values, context, beauty, vulnerability, ongoing multilateral commitments, and trade-offs. After struggling through these questions and finding that I could not, with any finality, resolve them — could not move from ethical questions to a moral code — I wondered if this was, indeed, the point.

The challenge in applying ethical principles to working with insects, and to eating them, is to articulate clear principles and guidelines while acknowledging the realities and uncertainties of living in a complex world. If an animal is in our care, and vulnerable, we wish to behave in such a way that we do not cause it pain or suffering. We have principles, and we care, but we need to keep ourselves from slipping into a kind of black-and-white moral code, which is a way of absolving ourselves of responsibility. It is not possible to live without also causing the deaths of others, whether we step on them in the path, eat plant foods that could have kept them alive, or eat them directly or inadvertently. Our pacts with all other living things play out in a complex web of multilingual nutritional, pheromonal, visual, and auditory conversations. Ethical interactions with insects are rife with unresolved, and perhaps unresolvable, questions. There are no ten commandments.

The best we can do is to pay careful attention, keep asking questions, and take responsibility for our actions.

# A LITTLE HELP

## *Regulating Entomophagy*

••••••••

*You say you've got a real solution.*
*Well, you know, we'd all love to see the plan.*

My mother-in-law, who grew up in China, used to say, "Locks are for gentlemen." The same holds true for regulations and trade agreements. If a desire for ethically grounded interactions with other species, and a dream of a more cooperative, congenial, sustainable, planetary existence are what draw people to entomophagy, then policies and regulations are the prenuptial documents that give that hopeful love some shape and safeguards. They are the locks on the doors.

We already have a scientific head start at assessing the nutrition issues involved in eating insects. The food safety issues have not, however, been well studied. In fact, much of what we think we know is based on analogy and inference from other classes of animals. Many of the issues are specific to insects and regions. A few are more general. At this point, the answer to the questions, "What are the food safety risks?" and "How can we best manage them?" are "We aren't sure" and "There is no *general* answer."

With that proviso, we can suggest some issues that entomophagy promoters would do well to consider. Some food-related illnesses are inherent in the food itself. For instance, allergic reactions to eating insects have been reported, as have cross-reactions between insects and related taxonomic groups, including shellfish. But whether one can generalize across arthropod groups remains an unsettled question. From other kinds of food allergies, we know that there is a complicated relationship between genetics, exposure rates, age at exposure, and a range of environmental factors. Asian countries report a higher prevalence of allergic reactions to shellfish (which are commonly eaten) and lower prevalence of reactions to peanuts and other nuts (which are less commonly consumed) than urbanized Westerners. China reports more than a thousand instances annually of anaphylactic responses to eating silkworm pupae. Again, this is likely a function of China's population size and higher rates of exposure to silkworms than in other countries.

Residues of industrial chemicals or heavy metals are almost impossible to get rid of once they are in a food. In this sense, they are similar to allergens; that is, they have become an inherent part of the food itself. If the best way to avoid food allergies is to avoid the foods altogether, then the best way to deal with metal and pesticide hazards is to find ways to prevent them from seeping into the food chain. With regard to insects, this usually means a shift from foraging to farming. Charlotte Payne and her colleagues identified the salt and manganese content of mopane caterpillars sold in South African markets as a concern. Other researchers have found elevated levels of copper, cadmium, and zinc in some of the caterpillars. In the early years of this century, chapulines (grasshoppers) exported from Oaxaca, Mexico, to southern California were demonstrated to have more than 300 times the lead levels deemed safe by the US Food and Drug Administration (FDA). Investigations in Oaxaca found widespread lead contamination from nearby mine tailings in soils, plants, and wild grasshoppers. However, the most important source of the lead in the chapulines appeared to be lead-glazed *chirmoleras*, which are small bowls used to grind spices.

Food safety issues are contextual. Protecting the safety and quality of food requires us to ask specific questions about specific insects. Where were they grown and processed, and under what ecological, social, and economic conditions? Mopane caterpillars in South Africa are as different from crickets in Ontario as chickens in Thailand are from cows in Switzerland.

As I have already recounted throughout this book, some insects, such as stink bugs, dung beetles, and mopane caterpillars, are inedible on an as-is basis either because of what they have eaten or because their bodies produce toxins. These insects need to be prepared properly before consumption. In parts of Africa, a diet that combines cassava, which contains cyanogenic glycosides, and silkworm larvae, which lack the sulfur-containing amino acids that are needed to detoxify the cyanogenic glycosides, may lead to thiamine deficiency; the people on this diet suffer from an acute ataxic syndrome.

Bacteriologically, farmed insects reportedly have similar risk profiles to other livestock, but this is an analogical inference, and there's very little research to back it up. Although the specific bacteria that cause spoilage and disease differ, similar methods are sometimes effective in managing risk from both groups. Most bacteriological and viral contaminants are destroyed by boiling the bugs, although spore-forming bacteria may sneak through the system. The Wageningen University food science group has determined that a mix of powdered, roasted mealworm larvae, flour, and water, when subjected to lactic acid fermentation, controlled bacterial contamination and improved the shelf life of the mealworm powder. If mopane caterpillars are degutted and dried, their shelf life is increases to almost a year. On the other hand, if not dried and stored properly, the mopane become susceptible to fungi, some of which are known to produce aflatoxins that have been associated with liver cancer.

How do we move ahead on entomophagic food safety issues? In most instances, we will need to work out safe ways to prepare and eat insects on a case-by-case basis. In other (non-insect) parts of the food industry, producers and processors often develop a Hazard Analysis Critical Control Point plan (HACCP, referred to as hassep for short). A HACCP plan involves identifying any points from farm to fork at which hazards can enter the food, and putting into place procedures to control or eliminate those hazards. For instance, living crickets may be contaminated with bacteria from floor litter, birds, workers, and visitors. However, if the buildings are secure and visitors are limited, the risks are minimized; then, if the crickets are cooked before being distributed to the public, for all intents and purposes any risks associated with the bacterial hazards are removed.

These kinds of plans depend on having some kind of monitoring system, a way to detect possible contaminants. While some companies have proposed using electronic sensors to detect bacteria and toxins, others have turned to the insects themselves for help. Scientists in the US and in the UK have trained bees and wasps to detect land mines, explosives, food toxins, plant odors, and fruit fly infestations before they are visually apparent. The parasitoid wasp *Microplitis croceipes* was shown to be ten times more sensitive to volatile chemicals than an electronic "nose."[105] Researchers are also working with fungi-eating beetles to determine if they can be used to detect pathogenic fungi on food.

For edible insects that are currently not commercialized — that is, they are grown, sold, or eaten in the "informal economy" — we should be consulting with those who have the most experience with harvesting and preparing those insects. If the bugs are inherently toxic but can be detoxified through processing in some way, then this needs to be done before someone eats them. And, as entomologist Alan Yen emphasizes, "the methods developed to make inedible species edible are an important intellectual property of the traditional societies that discovered them."

As I described earlier, Meeru Dhalwala introduced crickets into her restaurant menus in 2008 using a "soft" approach. The reactions I described earlier were those of customers, but restaurant inspectors also weighed in. Recounting how her restaurants had offered paratha (flatbread) that incorporated seasoned, roasted crickets, she said, "We were filling about two dozen orders in an evening. Everything was going great, until a reporter — we don't really know who — complained to the Vancouver health department. We had been so focused on the new dish that we neglected to notify them. That was entirely our mistake."

The cricket parathas were removed. After health authorities tested two[106] of the uncooked crickets for bacteria (as with any meat, the bacteria were there), there followed some polite back-and-forthing and then instructions from health inspectors to cooks 'for the handling and preparation of raw insect meats, at which point the parathas were put back on the menu. They remained there until the fall of 2011, at which time they were again removed. Dhalwala's plan, as she stated in a 2015 interview, was to reintroduce insects, but soft-peddle them until the customers got used to them.

If Canadians and Australians were soft-peddling their wares so as not to startle customers and health authorities, the latter being particularly susceptible to negative reactions, then the European entomophagists faced more complicated challenges. Some of the biggest ones stemmed from what might appear to be an entirely unrelated set of historical events.

In 1994, in the turbulent wake of the mad cow (BSE) storm across Europe, the European Commission banned the feeding of processed animal proteins (PAPs) from mammals to cattle, sheep, and goats. These proteins — essentially offal and other "waste materials" from slaughterhouses — had been added to the feeds of young pre-ruminants before their rumens developed, and thus before they could digest hay. The extra protein enabled them to grow faster and, in the long run, use less food. Before BSE, this recycling of slaughterhouse waste seemed to represent industrial ecology at its most efficient.

After scientists reported that BSE (bovine spongiform encephalopathy) and related transmissible spongiform encephalopathies (TSEs) were spread by ingestion of meat, especially nervous tissue, from infected animals, a ban on these PAPs made sense as a way to stop the spread of the disease.

In January 2001, afraid that there might be cross-contamination between foods intended for farm animals and those intended for dogs, cats, and other non-ruminants, the ban was extended, now excluding PAPs from all farmed animals. Fishmeal was the only exception. Nobody was thinking about insects. They had bigger issues on their minds: mass slaughter of cattle, trade barriers, and farmer suicides, for instance. They had to act — and be seen to act — decisively, if not always carefully.

In October 2015, most EU member states still officially prohibited the selling of insects as food, but there were reportedly no "precise" regulations in place. In the midst of this regulatory ambiguity, tales were circulating in the press of insect sightings in supermarkets in the Netherlands (insect burgers and nuggets), Belgium (burgers with buffalo worms), and the UK (bags of whole mealworms, crickets, and grasshoppers). In March 2015, Irma, a grocery store chain in Denmark and the second-oldest such chain in the world, announced that it would be selling edible insects. Two days after the insects hit the shelves, they were removed. "The goods are no longer for sale due to an unearthing of the authorities' views on selling these types of products," Irma spokesman Martin Hansen told a local radio station.

In Paris, in August of 2015, I heard rumors that authorities in some parts of Italy had pulled insects from a supermarket shelf, but I couldn't determine whether this was old news (a batch of silkworms from South Korea was rejected by border officials in 2012) or something new. The news from Italy was accompanied by an explanation that honey, royal jelly, propolis, and red dye from cochineal insects were the only officially allowable insect products on the EU market.

*Casu marzu*, the traditional Sardinian "delicacy" made by allowing fly maggots (*Piophila casei*) to crawl around in pecorino (sheep's milk) cheese, and *Milbenkäse*, a German sheep or goat cheese modified by mites, fell into a regulatory gray area and seemed to be allowed but only in certain jurisdictions.

In 2014, the Belgian Federal Food Safety Agency went out on a bureaucratic limb and approved a list of ten insects considered safe for human consumption. In 2015, the Belgians asserted that insects for human consumption "appear to offer great potential" as an alternative protein source and acknowledged that breeding and marketing insects was already being tolerated in some parts of the European Union.

The post-BSE food bans were a source of great frustration to Antoine Hubert and the European Association of Insect Producers. The bans meant that although Ynsect could argue that what it was doing made ecological and economic sense, and was in fact encouraged by FAO, they were legally prevented from making the logical step from research and prototype to commercial production. Hubert was also both impressed and annoyed at the way Enterra had used this situation to its own advantage, although Ynsect's own forays into pet foods and collaborative activities in Singapore and elsewhere demonstrated similarly deft business sense.

There is a long tradition of corporate leaders in the agri-food system railing against, and trying to thwart, government regulation. For those of us who have spent a few decades studying the global pandemics of foodborne diseases, however, it has often seemed as if the regulations were too little, too late, too fragmented, and too much reliant on the smiles and wiles of corporations whose reason for existence, after all, was not to "feed the world" but to make profits for owners and shareholders under the guise of feeding the world.

I would not color the new entrepreneurs at Ynsect, Enterra, and Entomo with the same brush we use on Tyson, Carghill, and McDonald's, but caution is not a bad thing when regulating a commodity as intimate as food.

The members of the European Association of Insect Producers were simply asking for clear regulations and a fair playing field, rules that would facilitate the entry of transformative, ecologically based technologies into the marketplace. Most of those getting into the insects-as-food business would concur with Afton Halloran and colleagues, who have bravely ventured into the chaotic jungles of insects-as-food regulations where — I am pretty sure, though I cannot prove it — there are mind parasites that kill brain cells on contact. Having emerged from those jungles, the authors — who must have had access to a vaccine against the mind parasites, as they emerged unscathed to tell the tale — assert that the "greatest barriers to the growth of an edible insect sector is the lack of all-inclusive legislation that governs the production, use and trade of insects as both food and animal feed."

That seems a fair characterization to me. The EU regulations, urgently put into place in the midst of a situation that was driven more by panic than by sober reflection had, at least in retrospect, probably overreached. In fairness, mechanisms to implement the kind of integrative, cross-departmental, interdisciplinary, extended-peer-review approaches that philosophers Silvio Funtowicz and Jerry Ravetz have called post-normal science (PNS), were still embryonic. PNS, Ravetz and Funtowicz argued, is required where "facts are uncertain, values in dispute, stakes high and decisions urgent." That certainly applied to the BSE situation.

Whatever one thinks of the original decisions, farmers — including those who raise insects — have had to live with them. The question now is how to proceed. Before looking a little more closely at the issues, I think we should consider why regulations are necessary at all. This consideration goes to the heart of what we think of as an ideal society, articulated by Jean-Jacques Rousseau as a social contract, in which citizens collectively constrain their desires in order to master their needs. In this context, the answer to why regulations are necessary goes something like this: if farmers are sending their goods into a system that distributes them around a region, a country, and, increasingly, the world, then those who are sitting down to share a meal with friends need some level of assurance that the food will not kill them, and that the lands and people on which those foods depend are being properly nurtured from generation to generation. Nurturing the lands and people includes fair wages for the farmers, reasonable assurance that the animals are well-treated, and protection against the spread of diseases in the animals and plants that the farmers are producing on our behalf.

If the farmer lives down the lane from me, or sells her produce through a local store, then I can check out her farm and get some assurance that way. That local system is built on trust (and keeping our wits about us). The more global systems are also built on trust, but, given the repeated betrayals by corporate leaders who are business-smart but biology-stupid, that trust needs to be expressed through regulations.

When parts of the food system are expanding rapidly, all kinds of people get into the business. Some opportunists see quick profits. Explosive growth in hamburgers, ready-to-eat salads, almonds-as-health-food, and chickens-as-lean-protein were all accompanied by epidemics of bacterial and viral diseases — think *E. coli*, *Salmonella*, bird flu. We have no reason to believe that insects are immune to spreading diseases; in fact, the entire focus of international and national regulations regarding insects to date has been on discovering how they can spread disease and how to prevent them from doing so. This creates serious challenges for introducing insects as human food.

A friend of mine who runs an open-air café and bakery received repeated visits from public health inspectors who were concerned about some flies in his kitchen. What if they landed on the pizza? They would be cooked, of course, and, although aesthetically disturbing, they would not be a public health hazard. But what if the pizza itself had an insect topping? What would this do to all the standard public health and food safety rules we have so diligently constructed over the past few decades?

Regulations governing agriculture and food have emerged from agriculture, health, and food-safety advocates — groups that don't often speak to each other. These regulations are furthermore a mixture of local, regional, and global. The regulatory situation is, to understate the case, a bit of a mess.

On August 10, 2015, the morning after having sampled larvae and locusts at le Festin Nu in Paris, I walked the few kilometers to the offices of the OIE (World Organisation for Animal Health), where I had an appointment with Brian Evans, Canada's former Chief Veterinary Officer, now the Deputy Director General of the organization. The OIE has devoted pretty much its entire existence to controlling or eradicating diseases and pests that affect animals that are not insects. From its perspective, insects have generally been viewed as a problem, not a solution. Since some of the countries who are members already trade insects across their borders, I wanted to ask Evans if the idea of insects as food animals — rather than as pests and disease vectors — was anywhere on their radar. Short answer: no. Longer answer: sort of.

The "sort of" applies to bees, which may provide precedents for regulating other insects. Beekeeping has become big business; this is mostly related to pollination services for monoculture crops (almond orchards, canola, and the like). Honey is often a sideline commodity for pollinators, but some parts of the honey business are worth hundreds of millions of dollars. One of the fastest-growing markets unrelated to pollination services is for manuka honey, produced by bees that get their nectar from *Leptospermum scoparium* and sold for its medicinal value.

Given the global size of the pollination and selected honey markets, bees have had better public relations departments at their disposal than other insects. The OIE does have a list of diseases affecting the western honey bee (*Apis mellifera*) and the eastern honey bee (*A. cerana*) in its Terrestrial Animal Health Code. This code requires countries to report occurrences of named diseases. These include bacterial diseases such as American (*Paenibacillus larvae*) and European (*Mellissococcus pluton*) foulbrood and various mites. Even for bees, however, the general system of regulations is a hodgepodge of national and local rules, usually depending on the goodwill of apiarists and their willingness to let their competitors and neighbors know if they are encountering disease problems. I will let you, dear reader, imagine how likely that is.

Beyond bees, there appears to have been mostly whispering, arm-twisting, horse-trading, and reports summarizing what little research there is, inferring about bugs from research on other animals. In 2015, even as FAO was encouraging the production and use of insects in food and feed, the OIE had no codes that would cover diseases such as the deadly cricket paralysis virus described by Australian researchers in 2000, the *Acheta domesticus* densovirus that decimated the North American cricket farmers in 2009, or *Linepithema humile virus 1* and deformed wing virus, associated with bee mortality and carried around the world by invasive Argentine ants.

FAO is concerned with food and agriculture. The OIE is concerned with diseases of animals, mostly farm livestock. So who deals with the food safety and public health issues? Theoretically, that would be the World Health Organization, through its Department of Food Safety; that department, however, has tended to focus on tracking epidemics of diseases as they occur in human populations and estimating the burden of foodborne diseases after the fact. One obvious place to direct global questions about insects as food would be to the Commission of the Codex Alimentarius, or "Food Code." Codex was established by FAO and the World Health Organization in 1963 to "develop harmonised international food standards, which protect consumer health and promote fair practices in food trade." Compliance with the standards, guidelines, and codes of practice recommended by the 187-member commission is voluntary, but they carry considerable force, being cited in the World Trade Organization agreements, for instance.

In 2012, at the seventeenth meeting of the Codex Coordinating Committee for Asia, Lao PDR, supported by Cambodia, Thailand, and Malaysia, proposed that food standards be developed for edible crickets. The proposal was not ratified. At the time I am writing this (late 2016), insects are mentioned in the Codex only insofar as there are allowable limits of insects or insect parts in other foods.

The Codex treatment of insects reflects the official practices of its national members and is in keeping with organizations such as the FDA, which establishes "maximum levels of natural or unavoidable defects in foods for human use that present no health hazard." The FDA publishes a list that includes such line items as apple butter, with five or more whole or equivalent insects (not counting mites, aphids, thrips, or scale insects) allowed per 100 grams of apple butter; frozen broccoli, with an average of 60 or more aphids and/or thrips and/or mites per 100 grams; and green coffee beans, where an average of 10 percent or more by count are insect-infested or insect-damaged. According to its website, "the FDA set these action levels because it is economically impractical to grow, harvest, or process raw products that are totally free of nonhazardous, naturally occurring, unavoidable defects. Products harmful to consumers are subject to regulatory action whether or not they exceed the action levels."[107] The FDA has a category of foods designated as GRAS (Generally Recognized as Safe); by late 2015, no insect-based foods had been approved under this designation. Nevertheless, the Massachusetts Department of Public Health did approve cricket chips for sale in grocery stores, and the FDA allows the sale of insects to people as food as long as the insects are farmed, not caught in the wild. Bottom line: the regulations are unclear.

In 2015, the tectonic regulatory plates shifted slightly. The European Food Safety Authority (EFSA) is an advisory group to the European Union. On October 8, 2015, the EFSA Scientific Committee published a "Risk Profile Related to Production and Consumption of Insects as Food and Feed," in which they concluded that "for both biological and chemical hazards, the specific production methods, the substrate used, the stage of harvest, the insect species and developmental stage, as well as the methods for further processing will all have an impact on the occurrence and levels of biological and chemical contaminants in food and feed products derived from insects. Hazards related to the environment are expected to be comparable to other animal production systems."[108]

Shortly afterwards, expecting the European Union to act on the report, Spanish authorities announced that they expected to lift the ban on marketing insects for human consumption. On November 19, 2015, the European Union put into place regulations with regard to new and innovative sources of food, which included insects, algae, and cloned meat. Nevertheless, the new regulations required that producers of such "novel foods" (and yes, there is a definition of those) submit a full dossier to the EU authorities demonstrating the benefits of their products.

Antoine Hubert, speaking on behalf of the International Platform for Insects as Food and Feed (IPIFF) said that those applying the new regulations on novel foods needed to look at ways to reduce the "administrative burden and the costs" for cash-strapped insect breeders.[109] When I had visited him in August 2015, his best-case scenario had been that the European Union would simply put insects into the exemption clause enjoyed by fishmeal. "Novel foods" was a kind of Plan B. Still, it opened the doors.

As we pondered the issues surrounding insect diseases, Brian Evans and I at the OIE shifted to the more general issue of what are called emerging infectious diseases. We agreed that everybody pretty much knew, based on decades of research, the driving forces that caused the emergence of the diseases that plagued humanity and our animals, and that the answers lay in addressing issues like land use, economic disparities, city design, and energy use and rethinking how food was produced and distributed. We also agreed that most global organizations would rather not think about these "political" causes of disease, preferring to focus on vaccines, drugs, and other short-term money-makers. The same issues apply to preventing and regulating disease spread through insect foods and feeds.

The regulatory wrangling in the European Union is about setting up food safety and quality assurance guidelines within their member countries. Nevertheless, it also has implications for trade, and for rules within the World Trade Organization and the Codex. These implications are very important for the businesses that are trying to find economically viable ways to get insects into the food system. I don't dispute any of that.

What has emerged quite clearly for me, however, is that no matter how clear and flexible the entomophagical regulatory framework we design is, it will never, by itself, achieve the promise of a globally sustainable, insect-based agriculture and food system. Necessary? Yes. Sufficient? No way. The regulations and policies are important in that they help us make explicit some of the issues we are collectively concerned about. They also push us to consider how we might deal with those really important, left-out things: gender, equity, animal welfare, compassion, and, perhaps, somewhere in there, that "love you forever" thing.

The policies and regulations are safety nets to guard against unscrupulous or ignorant producers and food sellers. They are the bureaucratic version of a prenup. But a marriage is not defined by a prenup.

ALL YOU NEED IS LOVE?
*Renegotiating the Human–Insect Contract*
• • • • • • •
*Are we happy just to dance?*

It is all very well to talk about loving insects and creating regulations, but in the world where we live, the human–bug relationship would fall into what Facebook calls the "complicated" category. To make things work, we are being asked to make a long-term commitment to a set of institutional arrangements even as we want to try to disrupt, or at least change, those arrangements. Although many issues related to insects in the food system can be managed by modifying food safety and disease management regulations, dealing with ecological issues is a much trickier business. Here we will find ourselves in a landscape characterized by a tension between regulatory fences and the open ranges of human and ecological possibility, between what we think we need legally and what we desire. This is also where the issues of

273

how we harvest insects raise their problematic heads and wagging fingers. For argument's sake, let us reconsider the general categories I suggested in the introduction — foraging in the wild, semi-management, and intensive farming — and see where the different entomophagical candidates fit.

Some insects are, and probably always will be, gathered opportunistically (cicadas, locusts) or seasonally (termites, grasshoppers, black flies). If one examines the issues that arose in the American and Madagascar locust plagues, the main issue preventing their uses as food was a lack of appropriate methods to harvest, process, preserve, and store them. Termites are probably never available in sufficient numbers to warrant new technologies. The periodical cicadas offer some interesting possibilities, and the success of a few enterprising companies such as the Anderson Design Group suggests that with the right harvesting, processing, and preserving technology, the staggered thirteen- and seventeen-year cycles could be translated into a rare and expensive treat from year to year. To my knowledge, no one has tried harvesting black flies, but there seem to be so many that surely there must be some opportunities for foodie entrepreneurs.

For insects that are semi-managed — that is, they could probably survive and maybe even thrive if our civilization collapses — what is lacking is a better understanding of how these insects interact with the landscapes in which they live and the infrastructures that can assure their access to sustainable, healthy food sources. These would include mopane worms and palm weevil larvae. We could also include honey bees here, but their status in human society is more ambiguous than that of the other two, and their history as a semi-domesticated species offers some lessons about the limits of domestication.

In several of the countries of southern Africa, including Namibia, South Africa, Botswana, and Zimbabwe, increased demand for mopane caterpillars has transformed them from a subsistence food into a valuable cash crop. South Africans themselves collect almost two tons of them every year, with a total value measured in tens of millions of US dollars. Botswana exports huge numbers of the caterpillars to South Africa, where many are packaged and sold or processed for livestock feed. Pretty much all of this is foraging of wild caterpillars, which has created serious concerns about sustainability.

The caterpillar population depends on rainfall and available mopane trees (*Colophospermum mopane*). Because of good prices and high demand, many collectors are harvesting greater volumes not only of the larvae, but even pupae, thus putting the future of the population of moths — and hence the sustainability of the harvest — at risk. This overharvesting is compounded by the cutting down of mopane trees for construction and firewood. In parts of Botswana, mopane caterpillars have disappeared altogether, and in Zimbabwe, armed gangs have reportedly attacked and robbed mopane foragers.

One response to overharvesting of the wild caterpillars is to intensify and intentionally manage production — that is, to shift from opportunistic foraging to farming. A four thousand-hectare woodlot can theoretically support almost 200 tons of caterpillars annually. In the Uukwaluudhi Conservancy in Namibia, traditional leaders have imposed restrictions on harvesting times; each harvester pays a fee to the community leaders, so that the leaders, who are presumably there to protect the long-term interests of the community, benefit. It's not exactly community engagement, but it's a start.

An intensification of the approach used by the Uukwaluudhi Conservancy could be developed, moving toward the kind of management one sees with palm weevil larvae. Such an approach requires creation of protected habitats and a good sense of the feed and space requirements of the insects at different stages of their life cycle. The unpredictability of moth population sizes and their uneven geographic distribution, combined with price and weather instability, make this a risky business. These are, after all, wild animals, totally dependent on the resilience of well-defined ecosystems. Recognizing this, many Thai farmers have taken up farming crickets, palm weevil, and mealworms.

Questions of social justice and ecological sustainability, as well as ethnic and gender equity, create further complications in shifts from foraging to farming. Globally, women and children tend to be the primary insect foragers. In Latin America and sub-Saharan Africa, women and children spend more of their time foraging than men, and insects are a higher proportion of their diet. Women in two Amazonian Yanomami communities, for instance, compensated for limited access to vertebrate protein, which was mostly eaten by the men, by eating more insects. Among the Tucano people in the northern Amazon, insects constituted about twice as high a percentage of animal protein for women than for men, who had greater access to fish and game. In sub-Saharan Africa, the rural women and children who have traditionally harvested mopane caterpillars to supplement their income and improve their nutritional status are being increasingly displaced by young, unemployed men. In general, these harvesters — men and women alike — are poor and, lacking market information and transport, get low prices.

In the past, when commercial enterprises and development

agencies have created programs to improve household nutrition and health in what used to be called "developing countries,"[110] and what I would now call insect-eating countries, the original rationale has been to help poor rural women and children. We see the same rationale being paraded out for raising insects. Often the people organizing these programs have had good technical and marketing skills; they know how to get chickens growing faster and staying healthier. These same experts, however, have often demonstrated a lack of social and ecological awareness. As some of these programs, such as those dealing with back-yard poultry, have been successfully scaled up into commercial money-making enterprises, men have taken control, leaving the women and children with little to show for all their hard work. Similar patterns are already occurring with insect-rearing. In Matt Broomfield's April 2016 *Motherboard* article on raising insects and the empowerment of women (which I referred to in the chapter on the "green hopes" of entomophagy) he acknowledges that "it is men who profit from the higher end of the insect market. The Zimbabwe study found that while women sell mopane worms at bazaars, bus terminuses and beer halls, the more lucrative wholesale trade remains the preserve of men. Women cannot access the infrastructure necessary to transport large quantities of worms cross-country. Moreover, as men have the capital to buy worms in bulk, on average they pay only 90 Zimbabwean dollars per kilogramme of worms, compared to the 160 Zimbabwean dollars paid by women."

Concerns about the environmental impacts of foraging have emerged in Southeast Asia and Japan as demand for edible insects has increased so rapidly. Insects are traditionally eaten in parts of Thailand, mostly in the northeast. In the past decade, exploding demand in urban and tourist areas has created severe environmental stresses. When the environmental impacts are within a country's national boundaries, that country can introduce regulations and management programs. When insects are imported, regulating gets more complicated.

In the world of entomophagy, imports are a short-term solution that externalizes environmental (and social) costs from consuming countries to producing countries. To meet consumer demands, Thailand now imports silkworm pupae, ground crickets, leaf-eating grasshoppers, mole crickets, and giant water bugs. Annually, wholesalers at Rong Kluea market on the Thai–Cambodian border import about 800 tons of edible insects from Cambodia, Myanmar, Lao PDR, and China. This includes about 270 tons of silkworm pupae from China (which are not foraged) and 170 tons of grasshoppers from Cambodia. The environmental impacts of these imports in the exporting countries remain unregulated and largely unexamined.[111]

In Japan, a number of factors complicate the picture. Charlotte Payne and her colleagues discovered that, although some 117 native species were traditionally eaten in Japan, the diversity and volume of insect consumption have dropped dramatically. Wasp larvae, grasshoppers, and silkworms are still eaten in measurable amounts in some parts of Japan, but native populations have been declining, as they are worldwide. Patterns of entomophagy in Japan, as elsewhere, are part of dietary changes in global popular culture, but they have been affected by pesticide use and industrial accidents such as that at Fukushima, on the one hand, and consumer demands for insects not just as food, but for entertainment and as pets on the other. Japan imports insects from Thailand, Korea, China, and New

Zealand. Without careful ecological management at the source, this is surely not sustainable.

Edward Hyams, in his landmark book *Animals in the Service of Man: 10,000 Years of Domestication*, discusses only three insects: silkworms, honey bees, and cochineal insects. He includes cochineal insects because, although the insects haven't been selected and bred, humans have cultivated host plants to specifically attract and feed the scale insects for the production of shellac and dye. Although they are an element in twenty-first-century food fights, and although they once fed a desert tribe in the Middle East, they are not generally considered part of the new entomophagy, and I shall not expand on their cultivation.

Silkworm production may offer some useful parallels for shifts from foraging to insect farming. Varieties of silk-producing moth larvae have been domesticated in China since about the third millennium BCE, in India from about a thousand years later, and on the Greek Island of Cos from about the fifth or fourth century BCE. For economic and political reasons, Chinese silkworm production, based on the breeding and selection of *Bombyx mori* moths, has historically dominated global production. Like the weevils and mopane caterpillars, the moth larvae used in conventional silk production require food from a particular tree and hence are dependent on the maintenance and/or cultivation of that tree. Over a period of thirty to forty days, caterpillars grow to ten thousand times their hatching weight; during that time, thirty grams of the caterpillars will eat through more than a ton of freshly picked mulberry leaves, so that, without domestication, the industry would cause serious ecological damage.

The particular moth's association with a particular tree and eco-cultural history in China makes it an interesting case study to look at when considering several other edible insects, such as mopane caterpillars and palm weevils. Silk production in other parts of the world has recognized and respected the cultural origins of the cultivation and ecological features of the moth, so that the benefits are more widely distributed than if sericulture were treated as a generic insect production system in a way similar to, say, chicken production. The problems of trying to introduce sericulture without understanding its ecological context are well illustrated by the initial attempts to make the United States into a great silk-producing country in the nineteenth century. According to entomologist Gilbert Waldbauer, a resolution put forward at an 1842 meeting of the New England Silk Convention read, "Resolved: that, inasmuch as in America and China the mulberry tree is found in the native forests, it is manifest indication of Divine Providence, that this country, as well as China, was designed to be a great silk growing country."[112] What the early American silk enthusiasts did not understand was that *Bombyx mori* much preferred China's white mulberry trees (*Morus alba*) over America's red ones (*Morus rubra*). As Waldbauer adds, "Perhaps the author of the resolution misread Divine Providence because his botany was weak." One challenge of intensifying mopane caterpillar, palm weevil, and cricket production will be to ensure that understandings of biology, culture, and the relationships between them are strong.

It appears in general that as the demand for edible insects increases, management of protected areas and sanctuaries might be the most sustainable option for foraging in the wild, if careful attention is paid to social and economic relationships during this transition. I was encouraged when the reports from the UK's first public Insects for Food and Feed Conference (in 2015) explicitly addressed the challenges of trying to solve wicked problems in a complex world. Charlotte Payne, Andrew Muller, Joshua Evans, and Rebecca Roberts, for instance, encouraged participants to repoliticize the insect-eating movement, contextualizing it in the inequitable, complex agri-food system we inhabit. Others raised questions of who we mean when we talk about "us" feeding "them." Darja Doberman, a graduate student — in my experience, they are usually the leading-edge scholars — proposed that in parts of sub-Saharan Africa, crickets could be grown on millet husks left over from brewing beer, and then millet-based foods could be fortified with cricket flour. This would seem to represent a strategy that respects ecological concerns, local culture, public health, and nutritional science.

For insects that are intensively farmed, such as crickets, mealworms, soldier flies, and perhaps silkworms, we can probably adapt research, management protocols, and regulations already in place for other intensively managed animal species. A lot of the research into livestock agriculture in the past few decades has focused on efficient use of feed resources, FCRs, and managing environmental pollution, and insect farming can benefit from some of that work. Nevertheless, there are additional complications. For one thing, they are often being introduced as a solution to problems identified in other livestock agricultural activities, such as greenhouse gas emissions, pollution, depletion of fish stocks, and clearing of the rainforest to grow soybeans. As one would expect with any wicked problem, the "insect solution" creates new problems even as it solves old ones. However, with some careful attention to environmental, gender, and economic issues, farming them can provide a useful, "soft" entrance for insects onto our plates.

If silkworm cultivation offers one way to think about how we manage our relationships with edible or otherwise useful insects in transitions to domestication, honey bees, with their ambiguous status as both wild and intensively managed, offer both helpful analogies and serious warnings about the limits of domestication and the unintended consequences of new technologies.

In terms of the links between the ecological niches in which bees thrive and value-added product diversity, as well as the introduction of innovative technologies, the history of human–bee relationships is instructive. Farming food animals and crops is a high-risk business, and one survival strategy is to develop multiple, diverse, and value-added products. In dairy farming, this means yogurt, many different kinds of milk and cheeses, veal (from bull calves), and hamburger (from old cows). And just as not all milk is created equal — that from Jersey cows has a higher fat content and is therefore valued differently from Holstein milk — not all honey is alike in value.

Some of the value of the honey is related to the specific varieties of pollen and nectar eaten by the bees. The New Zealand (manuka) teatree *Leptospermum Scoparium*, for instance, is the basis of a billion-dollar industry due to the medicinal qualities of the honey produced by the bees that harvest its nectar. Rich in the antibacterial chemical methylglyoxal, manuka honey sells for as much as ten times as non-manuka honey. As one might imagine, this has spawned a rush by other countries to find their own sources of methylglyoxal-rich honey. While people would like to be able to feed bees on specific crops to reap the economic benefits, the bees, like all of us, need diversity in their diet.

The pollen that the bees eat, which has protein content that varies by plant and by season — from under 4 percent to over 40 percent — is necessary for physiological development. Hence the *volumes* and *varieties* of pollen bees have access to are as important as the fact that there are flowers around.

This need for a diversity of possible pollen and nectar inputs is a key factor in how bees have become essential to agriculture, and also explains in part why they have not adapted well to industrial monocultural agriculture. Honey bees thrive as a domesticated species only if their human caretakers provide access to a diversity of protein-rich pollen. One of the most dramatic, and ultimately problematic, innovations in the history of modern beekeeping occurred in 1851, in Philadelphia. Reverend L.L. Langstroth designed a hive composed of stacked boxes that contained move-able, framed combs with fixed wire frames and plastic foundation sheets to standardize the cell size and comb structure. Langstroth hives have separate compartments that exclude the queen and hence are full of pure honey, rather than honey mixed with brood. New boxes are added to the top as the hive fills up. This design enabled easy automation and set the stage for the entry of bee-keeping into its current role as a key player in the industrialization of agriculture. The bees do all right in Langstroth hives, but the hives were designed for the convenience of people, not for the well-being of bees. Bees may have trouble accessing their own honey stores in a Langstroth, and can starve during droughts or in winter even though there is honey close by in the hive.

Langstroth hives have enabled a multimillion-dollar migra-tory pollination industry in many industrialized countries; thou-sands of hives are loaded up on trucks and moved from one monoculture site (almonds, cherries, blueberries, canola) to another. If they are not trucked elsewhere after the monoculture has flowered, the bees are left trying to survive in what to them is a virtual desert. Once in the system, the pollinators are in a bind: truck on to the next place or go home and look for more diverse pollen sources. Large-scale pollination businesses have been reported to have 20 percent normal losses, and up to twice that percentage in those where colony collapse disorder occurs. All beekeepers experience deaths and losses as bees starve, freeze, or suffer from a variety of diseases. Nevertheless, the high rates of hive loss in recent years is yet another sign — like salmonellosis in raw meats — of what we have come to consider "normal" for industrial agriculture.

Coming back to Cahill's idea that renewal comes from the margins, where can we look for alternatives to the current, Langstroth-dominated industry while still acknowledging the contractual agreement we have with honey bees? Adam Gopnik, in a BBC commentary on the history of how Europeans viewed the queens and kings of honey bees, recounts the story of Charles Butler's 1609 observation and announcement that a beehive was "the feminine monarchy." Gopnik concludes his piece by saying that "one moral of the tale of the bees is, of course, always to trust the Butlers rather than the Aristotles of the world. Trust the man who sees the bees instead of the old Greek philosopher who just had opinions about them."[113]

With that in the back of my mind, I decided to consult my very own in-family "man who sees the bees," my son Matthew. On his website, he argues that his use of Warré hives "attempts to emulate the way bees build inside a hollow tree.... New, empty boxes are added to the bottom of the hive, so the brood are always protected by an insulation layer above. The queen always has the option of laying her brood in new comb, and when it hatches out, the bees can put honey into that comb. The hive has an insulation box on top to keep the temperature more stable inside the hive, and the walls of the hive are made of thicker wood for better insulation. In winter the bees are free to move up or down the hive to find honey. Fewer stresses on the bees mean they can deal with other things (disease, pests, seasonal variability, environmental toxicity). The entire structure is well suited to staying in one place."

Matt has followed the usual Warré apiarist practice of adding his hand-crafted boxes to the bottom of the hive (nadiring), which requires heavy lifting. Also, he notes that honey harvest is lower than with a Langstroth, although the selling price for such artisanal honey is higher than for commercial products from larger apiaries.

"In the end," says Matt, "I get better quality honey and happier, healthier bees. The benefits far outweigh the costs in my opinion!"

As we consider expanding, diversifying, and more carefully managing insects in our agriculture and food systems, can we learn from the unintended consequences of innovations in beekeeping? Can we manage crickets, mealworms, palm weevils, and mopane worms in ways that respect our relationships, including the benefits we share in those relationships? One salient lesson we can learn from the history of beekeeping is that sustainably and ethically managing mopane caterpillars, crickets, mealworms, and palm weevils will require us to pay attention to a whole range of social and ecological conditions and interactions that go far beyond FCRs, shelf life, and consumer attitudes. Once we step out of the laboratory, what previously appeared to be observable, independent "facts" become malleable conversants speaking languages that range from pheromones to singing, and from magnetism to visual perception. The laboratory facts are but a few words contributing to dynamically changing networks of communities.

The idea that technology experts and scientists who wish to make a difference in the world need to work with the communities that are the intended beneficiaries of their altruistic efforts is not new, but it seems to need repeating, reinforcing, and recharacterizing on a regular basis. In part, this reiteration is necessary because there are some deep tensions in this proposed relationship between the scientific community and the world at large. Scientists want to make general claims about the world: smoking is bad, crickets are good, pesticides are bad for health, pesticides are necessary for good nutrition and health, commercially produced mopane caterpillars will solve protein malnutrition, insects will provide global food security. But, as even the National Science Foundation in the United States asserted in a 2003 report, the kinds of questions raised by interactions between ecology and human well-being often require a "place-based" science, and, unlike laws of gravity or the speed of light, ecological and health-related truths are not universal. Scientists and the communities they work with have very different agendas and perspectives. Communities are internally heterogeneous, with their own historically based power structures, gender dynamics, economic activities, ecological constraints, and a mix of desires and goals that are often not well articulated.

This tension between what we think of as normal, puzzle-based science and the complexities of the universe we inhabit is not new, nor is it unique to entomophagy. We now have several decades of theoretical and applied scholarly work to guide us through the mess in some reasonable fashion.

There are ways of working with and through these dilemmas

and tensions. In the 1990s, in an attempt to balance the natural world and the human-constructed one, some of us pictured this kind of integration in what we called the "Butterfly Model of Health," with the biophysical environment as one wing of the butterfly and the socioeconomic environment as the other.[114]A picture is a useful heuristic, a don't-forget-this checklist. In practice, we were working from a scientific base of what we called the "new science," or "post-normal science," which I referred to when talking about BSE. I would direct any applied scientists and scholars in the entomophagy field to explore that literature. The bottom line is that, both from a scholarly point of view (to get good-quality evidence) and from an applied perspective (to use that evidence to improve the world in some way), the peer group needs to be expanded to include a wide range of people who have experience and information on a subject, and others who will be affected by how one acts on the information. Such approaches, in which those who live in the areas being protected or managed are integrally involved in creating, implementing, and assessing adaptive plans, have a documented history of effectiveness in achieving multiple, interacting goals. These goals now include health and nutrition, ecological resilience and biodiversity conservation, ethics, and welfare. Based on the mixed legacy of previous attempts at improving human well-being, we can add economic and political sustainability, ethnic and gender equity, and the more challenging issues raised by recent attempts to integrate human, animal, and ecosystem health. Examples of these integrative approaches include EcoHealth, One Health, and Resilience.

Recently, Emily Yates-Doerr, an anthropologist at the University of Amsterdam, raised many of these issues specifically with regard to how entomophagy-related research is structured and its results promoted. Her 2015 article, "The World in a Box? Food Security, Edible Insects, and 'One World, One Health' Collaboration," examines the tensions between the way laboratory scientists think about problems (linear, boxable, replicable, exportable anywhere in the world) and the way eating habits and taste preferences emerge from complex local, historical, cultural, and eco-social dynamics. In my quarter-century of teaching the epidemiology of foodborne and waterborne diseases, I found that I had to repeat, year after year, that people do not *only* eat for nutrition. We eat particular foods prepared in certain ways because of the quirks of our histories, for enjoyment, and as a source of identity.

As Yates-Doerr concludes in her research paper, "The results of the scientists' research suggest that for any particular food security initiative to succeed, more than 'One World' or one form of health must be incorporated into the research framework. To impact the food supply of 'the world,' it is necessary to attend to many different worlds."[115]

Having been around the One World–One Health–EcoHealth block more times than I care to remember, I agree with Yates-Doerr, and I would extend that way of reframing to life beyond research. The oneness of the world in which we live emerges from, and is only possible because of, complex relationships among "millions and millions" of diverse organisms, people, landscapes, and cultures. Our challenge as entomophagists — and as humans — is to envision the oneness even as we nurture the diversity.

WE WERE TALKING

*Where Is This Going?*

• • • • • • •

*Crickets and this guy with diamonds?*

The flight from Toronto to London, followed by the Heathrow Express and then the crowded metro to the small hotel near Regent's Park left me aching and a bit queasy. It was almost noon. I had a 1:30 lunch reservation at the Archipelago Restaurant,[116] which is known for its exotic and eclectic menu, including several dishes that feature insects. I figured the hour-long walk in the bright sun across the park and some light bug dishes would help me get my feet back on the ground.

Entering Archipelago is a bit like walking into a nineteenth-century curiosity shop, chockablock with peacock feathers, Buddhas, and wooden Indonesian wayang golek puppets, a collage of greens, reds, pinks, and browns, wood, glass, cloth, and brass.

After telling them the password they had given me to hold my reservation, I was offered a table near the window with a glass Buddha with a reddish translucent body, a gray head, and a crown of golden curls for my dinner companion. He didn't say much but seemed to enjoy the ambiance. There was one other person in the restaurant, a clean-cut thirtyish American studying archaeology in England. He too was trying all the bug dishes, announcing that he was probably the only adventurous eater in the small Michigan town where he'd grown up. I asked him how he liked the insect dishes, and he was enthusiastic.

My lunch consisted of Summer Nights (Pan fried chermoula crickets, quinoa, spinach, and dried fruit), Love-Bug Salad (Baby greens with an accompanying dish of zingy, crunchy mealworms fried in olive oil, chilis, lemon grass, and garlic), Bushman's Cavi-Err (Caramel mealworms, blinis, coconut cream, and vodka jelly), Medieval Hive (Brown butter ice cream, honey and butter caramel sauce, and a baby bee drone), and Chocolate-Covered Locusts (white, milk, and dark), served with a small glass of sweet white wine. The insects were integrated into the dishes, adding crunchy texture and subtle flavors.

I chatted with the man serving me, an Australian who had once been an event organizer at the Sydney Opera House and more recently had led tours around the UK and Europe. Himself an eclectic world traveler, he fit right in. The restaurant had been started by a guy from South Africa, who had seen a need for the kinds of "exotic" meats on offer — zebra, crocodile, pythons — in his adopted hometown. Insects had been on the menu from the get-go, so Archipelago was not part of the "new wave."

Archipelago had more insect-based sweets on the menu than I have seen elsewhere, and with their mix of chewy and crunchy and the understated flavors of honey, caramel, nuts (in the Bushman's Cavi-Err), and chocolate, they surprised and delighted me. The insects were clearly present, but not in my face; the food had a more relaxed feel than the insect dishes at le Festin Nu or Uchiyama-san's Tokyo street theatre, and seemed more eclectically "normal" than the hip conversation starters at Public in Brisbane or the light sprinklings of insects at Billy Kwong in Sydney.

After lunch, I thought about one of the restaurants I had not visited on my 2015 travels — Noma in Copenhagen, where Chef René Redzepi had led, cajoled, and tyrannized his kitchen staff to two Michelin stars and several "Best Restaurant in the World" awards. As I mentioned earlier, Noma had also been hailed as a champion of entomophagy. Then, in 2016, after watching Pierre Deschamps's dramatic and revealing documentary *Noma: My Perfect Storm*, I went back to look more carefully at the stories under the headlines. Although Redzepi had certainly introduced insects onto his menu, promoting entomophagy was not a large part of his agenda. As the champion of "Nordic cuisine," Redzepi's mission was to get chefs to discover edible — and delicious — plants and animals in their local ecosystems, Nordic or otherwise. The intent was to have chefs and diners alike become more aware of the natural ecosystems in which they lived.[117]

Suddenly my experience at Billy Kwong in Sydney made sense. They were following the hot, "rock-star" global chef, downplaying insects and emphasizing locally and seasonally available produce. All good stuff, but not about promoting insects on the plate. Just normalizing them as part of a diverse food palate. This also resonated with what "lumberjack" Daesuke-san in Kushihara, Japan, had told me. They ate local foods and, yes, occasionally hunted and consumed hornets, but that did not define who they were. In October 2016, I asked Vancouver chef Meeru Dhalwala how the reintroduction of insects was coming along. It was a continuing challenge, she said, and they'd made no further moves to do so at their restaurants. In fact, she had recently spoken at a "Future of Food" conference, where her topic had been "Insects, Seaweed and In-vitro meats." Insects, then, as part of a diverse response to generating food options in a crowded, resource-challenged planet.

When I consider the role of insects in the "Western" cuisine of the future, I recall how my three-year-old granddaughter, when I emptied a few packets from Entomo Farms into small bowls on the table in front of her, didn't hesitate for a second. She just gobbled them up and then tipped the bowls back to get the last few crumbs. I asked her which she preferred and she said, again without hesitation, that she liked the mealworms, because the legs didn't get caught in her teeth. And I can still picture one of my grandsons, who, also about three, dug into his Christmas stocking and pulled out two snack packs from Entomo Farms. His response? "Crickets! Yummy! Mealworms! Wow!"

All things considered, I'm not really worried about whether or not insects will become a more central part of the human diet, whether on earth or out into space.[118] We will get over the wave of insects as a hot new trend among foodies. I suspect that, on the human consumption side, restaurants like Public and Archipelago will become the norm, and, on the production side, we will see a wide diversity, from Ynsect to Entomo and Enterra. Over the next few decades, billions more of us will deliberately eat insects, and by deliberately I mean we'll eat them by choice, and not simply in the form of the insect bits we already devour in our coffee and bagels, lentils and tea, or burgers and ketchup. Not everyone will eat them, and some will only eat them from time to time, as part of a varied diet. I occasionally eat shrimp when I travel, but only if I am near the ocean, and they have not traveled across continents to my plate. To me, they are neither adventure eating, nor new, amazing, weird, or world-saving — all those labels that are applied to eating insects. At home, I don't eat them. We live far from the ocean, and my wife is allergic to shrimp and shellfish. I eat insects when I share a meal or a snack with other insect-eating humans. As we move further into the twenty-first century, some people will continue to eat insects as important sources of protein, as part of a diverse diet in a complex world, or as a way to demonstrate bravado; in other situations, insects will be used to enhance the nutrition of people at risk, such as those in refugee camps; still others will eat them as ingredients in a frugal life.

I am delighted with the prospect that some insects will show up as an option in our supermarkets and on our plates. I am more concerned about how that process happens, and the unintended consequences of bungling that process. The most active leaders in the new entomophagy movement are helping us to rethink the ways we as humans feed ourselves and to ground what we do more completely in ecologically sound practices. Even as some are working to close loops in the agri-food system, recycling waste and replacing fishmeal and soy with insect products, others are finding ways to reinvent our agri-food practices completely.

During the historic shift from hunting and gathering to agriculture, our ancestors allowed the animals now so entrenched in our food system — cattle, pigs, sheep, even fish — to creep in, almost without thinking. This is the first time in recent history that humanity is faced with the possibility of making some conscious decisions, based on the best available information we have, about what sorts of animals and practices might help provision us in a heavily populated world. If we see it merely as a technical issue, or a way to add yet another item to our diet, or even a way to reduce our footprint, we will have bungled this once-in-a-millennium opportunity.

Torres Strait islander Kerry Arabena, in a vision that transcends and integrates political and ecological history, argues that we are all indigenous to the universe.[119] Our imaginative and physical renewal will come from rediscovering this sense of indigeneity, drawing on a rich mixture of indigenous and local knowledges and the various sciences, experiences, and explorations of eco-social complexity that characterize Western scholarship. From this perspective, renewal can come from a kind of deep ecological understanding and rediscovery of the millions of arthropods who have made us and sustained us.

This, then, is my aspiration for the new entomophagy movement: not just that we will put yet another item on our plates, but that in looking at the insect world more closely, we will see the world, and imagine ourselves, in a new way. Perhaps, in exploring the possibilities of insects as food, we will discover a more complex understanding of ourselves.

I would like to see insect-eating as a normal culinary option. I would like to see it as a way to open our eyes to the rich array of human cultures and diverse ecosystems we inhabit, to reimagine our place among the millions and millions of other animals we share it with, and to open collective eyes to our common indigeneity in this puzzling and stunningly mystifying universe.

Such an awesome planet. So little time. My advice to aspiring entomophagists is to get out there and learn from the experts: the people who live where they eat, the attentive ones, the ones who see beyond commodities, the ones who care. Find the voices. Listen to them. Share their stories across cultures. Tell them, and retell them. Redefine normal. You can still make a difference.

# PART VII. REVOLUTION 9

In the old stories, perhaps a tale by Graham Greene or Somerset Maugham, after the intimacies of eating, or sex — which are, after all, similar, just with different partners — the man or the woman, or both, would have a smoke, or a brandy, or both, pondering the meaning of what had just transpired. Since the nineteenth century, insects have caused Europeans like Darwin to question their most fundamental religious beliefs. Can we eat crickets and mealworms and talk about the meaning of life? Can insects on the plate not just help us to live longer, but also help us learn how to live well? As John Lennon proposed, let us imagine heaven here on earth.

IMAGINE

*Beetles, Entomophagy,*
*and the Meaning of Life*

• • • • • • •

*"Imagine there's no heaven."*
—John Lennon

*"The slogan of Hell: Eat or be eaten.*
*The slogan of Heaven: Eat and be eaten."*
— W.H. Auden, *A Certain World: A Commonplace Book*

About a millennium BCE, when managing honey bees already
had a long history in the Middle East, and prophets were eating
locust proteins to supplement the sweets they pillaged from bees,
a fierce, violent, emotionally volatile bard announced that "the
heavens declare the glory of God; and the firmament sheweth
his handywork" (Psalm 19:1). It may have been the bravado of
a bloodthirsty warrior, sure that the Old Man had his back, but
it was a sentiment with which many European naturalists and
natural philosophers would later concur. In 1738, for instance,

Friedrich Christian Lesser — a physician and member of the German Academy of Natural Scientists — published *Insecto-theology: Or a Demonstration of the Being and Perfections of God, from a Consideration of the Structure and Economy of Insects.*

This same sentiment is reflected today by even the most aggressively atheist neo-Darwinists, who proclaim that the diversity of insect species reflects the genius of nature itself. The general argument is that, in understanding nature, we are not only pursuing the ancient Delphic maxim that we should "know ourselves"; we are also somehow bettering ourselves socially and morally. As Dr. Samuel Johnson famously declared to his biographer, James Boswell, after rescuing a bug that had crawled on his nose, "There is nothing, sir, too little for so little a creature as man. It is by studying little things that we attain the great art of having as little misery and as much happiness as possible."

All this sounds good, but really, in terms of everyday life, what does it mean? European naturalists, as well as the teachers and camp counselors who guided my childhood, saw nature as an illustration, a lesson. For them, the idea that nature could be understood and valued on its own terms, like the notion that art could be valued for its own sake, was a foreign concept; nature was seen to provide an almost endless storeroom of moral lessons, food, and materials for building stuff. The idea that nature must be *for* something is deeply embedded in the twenty-first-century perspective that runs through much of the sustainable development literature, as well as magazines such as *The Economist*, which is less well known for its strong environmental stances. The twenty-first century version of *insecto-theology* asserts that the biosphere is best understood as a provider of ecosystem *services*, like water, food, and entertainment — the natural counterpart of a laundry service for diapers or hospital scrubs. Or even a platform providing open-source software that generates metaphors and philosophical ideas.

Recent proponents of entomophagy have made similar arguments. Entomophagy, we are informed, is a way to reduce the size of our ecological footprint, reduce greenhouse gas emissions, eat in a sustainably healthy fashion, and create an eco-friendly society. At one level, I find this hopeful, interesting, and even, some days, exciting. At a deeper level, I am troubled. As Nobel prize–winning scientist Joshua Lederberg wrote, in an introduction to *Haldane's Daedalus Revisited*, "Above all science is bereft of deontology: it cannot tell why one should be interested in science or anything else."[120]

Entomophagists imply that we *should* want to develop sustainable ways of living in the natural world because we *care* about it. I agree, but I ask myself, *why* do I agree? Caring, as I argued earlier, is the basis for an ethically grounded response to insect suffering. But if science cannot provide an answer as to why we should be interested in the world to begin with, where can we begin to look for a reason as to why we should even begin to care? Why should I care if entomophagy is, in Daniella Martin's words, "the last great hope to save the planet"? Who cares if we kill each other, destroy landscapes, and drive species to extinction? The planet will someday — billions of years into the future, or tomorrow — be extinguished. So we trash the place and leave early: why care?

Some evolutionary biologists argue that we have a common-sense notion that we should be good to each other in some vague way. But again I ask, why? Because this will enable us to win the evolutionary race to out-breed everyone else? We've been there.

Done that. Some of us are past breeding age, and are relieved and happy to have finally arrived here.

There is an element of that attitude that nature is useful — in this case useful as an illustration — in the (possibly apocryphal) story about J.B.S. Haldane, who, so the tale goes, found himself in the company of a group of theologians. On being asked what one could conclude as to the nature of the Creator from studying his creation, Haldane is said to have answered, "an inordinate fondness for beetles."

Following in the spirit of Haldane, one reason to write a book about eating beetles is to understand the mind of an alleged Creator and then, in some naturalist version of a shared feast such as communion or Passover or Eid al-Fitr, to eat him (or according to your metaphorical tastes, her, or the great genderless All-Encompassing Being).

I am going to risk something here, but in considering entomophagy as a narrative thread through the evolution of life's diversity and messy connectedness, it would only be the timid and small-minded who would retreat to the Cartesian laboratory, in which *how* is consistently confused with *why*. At some point, the answers to a five-year-old's persistent refrain of *why?* — because of DNA, because of gravity, because of pheromones — begin to ring hollow, and one is left with saying, well, *because that's how it is!* Which of course is a recognition of some kind of failure — whether of intellect, courage, or imagination I am still unsure.

What, after this quest through the webs and tunnels of the world, can we now say on the subject of a universally creative being who has an inordinate fondness for beetles? Does speaking of *meaning* after pondering the material structures and mechanisms of insects, evolution, and entomophagy not seem excessively presumptuous, in the same class of arrogance as the Dawkinites' self-description as "bright ones"? Maybe. I'm going to call it late-stage bravado. Like the Canadian novelist Margaret Laurence, I believe that, as we get older, we should become more radical, more venturesome, less tolerant of the frass of intellectual timidity that plagues so much of modern life. I think I'm in reasonably good company here.

"What is it that breathes fire into the equations and makes a universe for [humans] to describe?" asks the physicist Stephen Hawking in *A Brief History of Time*. "The usual approach of science of constructing a mathematical model cannot answer the questions of why there should be a universe for the model to describe. Why does the universe go to all the bother of existing?" Hawking concludes his book by announcing that "if we do discover a complete theory, it should in time be understandable in broad principle by everyone, not just a few scientists. Then we shall all, philosophers, scientists, and just ordinary people, be able to take part in the discussion of the question of why it is that we and the universe exist. If we find the answer to that, it would be the ultimate triumph of human reason — for then we would know the mind of God."[121]

Hawking's mistake, conditioned by decades of thinking like a physicist, is that he is expecting theory to provide the fire, but that's exactly the problem, isn't it? The fire is beyond the theory, just as reality is beyond language. To imagine the fire behind the mysterious forces of gravity, spaces between quirks and quarks, stars, planets, and black holes, does not lend itself to theoretical constructs that lead to prediction. The question is how we can begin to understand ourselves as one rare animal among millions of others. Or, perhaps more accurately, according to those, like evolutionary theorist Lynn Margulis, who say we have evolved from bacteria and are in fact complex communities of collaborating bacteria, we are one animal composed of trillions of other, smaller organisms.

As with so many important things in life, we don't really have a good language to talk about this. In the same introduction to *Haldane's Daedalus Revisited*, to which I referred above, Lederberg declared that biology "is already so fact laden that it is in danger of being bogged down awaiting advances in logic and linguistics to ease the integration of the particulars." And as Albert Einstein famously argued, "We can't solve problems using the same kind of thinking that we used when we created them." Yet all our languages carry with them the "same kinds of thinking," the intellectual constraints, social baggage, biases, and blinders of the cultures from which they emerged.

In part, then, this quest for meaning is a search for a language. Some have proposed English, that omniglot, ever-syncretic lingo of a small island, as a possibility. Others might consider other religious or politically important languages: Latin, perhaps, or Arabic, or Chinese, or Russian. Early Cartesians dreamed that science might provide a universally understood sort of Esperanto. Others have proposed mathematics as the universal language, which works if you are a mathematician or a physicist. But none of these encompass Hawking's dream of a language that enables "philosophers, scientists, and just ordinary people be able to take part in the discussion of the question of why."

The traditional language often uses shorthand names, often connected to stories, each with their own baggage, as a starting (and, too often, finishing) point. There are hundreds of these.

The more familiar ones would include God, Allah, Yahweh, Brahma, and Ahura Mazda. Others have attempted to escape the cultural rootedness of names and refer to characteristics: light, goodness, love, fire, the force. All these names aspire to recognize what lies behind the words, the invisible baggage carrier. Indeed, we all need constant reminding that the words we use are not the things to which they refer. The names are shorthand ways to communicate, and that's something that, as humans, we need to live with. All language is metaphoric and I, for one, celebrate that. The problem arises when the scholarly authors, oblivious to the cultural baggage borne by their supposedly neutral descriptions, make definitive claims.

Darwin saw the behaviors of some parasitic wasps as a reason to abandon a belief in the Victorian version of God. In an 1860 letter to the American naturalist Asa Gray, Darwin wrote, "I own that I cannot see as plainly as others do, and as I should wish to do, evidence of design and beneficence on all sides of us. There seems to me too much misery in the world. I cannot persuade myself that a beneficent and omnipotent God would have designedly created the Ichneumonidae with the express intention of their feeding within the living bodies of Caterpillars, or that a cat should play with mice." Darwin's abandonment of "God" — like Hawking's desire to understand the "mind of God" — is not about gods in general. Darwin is rejecting a particular sort of god, the curmudgeonly Old Man so beloved by politically and economically powerful patriarchs and kings and businessmen because he's clearly on their side, and by self-styled know-it-alls and revolutionaries because he's such an easy target. Hawking is talking about something else. But what?

Harvard paleontologist Stephen Jay Gould, an acute observer of the natural and cultural complexity within which we and our sciences have emerged, has commented (with regard to the apparent moral conundrums raised by parasitic wasps) that we "seem to be caught in the mythic structures of our own cultural sagas, quite unable, even in our basic descriptions, to use any other language than the metaphors of battle and conquest. We cannot render this corner of natural history as anything but story, combining the themes of grim horror and fascination and usually ending not so much with pity for the caterpillar as with admiration for the efficiency of the ichneumon."[122]

The head-banging contradiction at the heart of this is that on the one hand, we, who consider ourselves at the very least to be reasonable and sometimes rational, are evolution made conscious of itself, who suddenly understand the truth of how we came to be. On the other hand, the truth, as we understand it, is that we are here through processes of random mutations interacting with natural and human-devised selection pressures and disasters, from incoming comets and earthquakes to polluted waters and desertified grasslands. And the sole relevant outcome of this process is that one's offspring live long enough to reproduce. We are here now, us waterbags of chemicals, microbes, and bugs, anxious cucumbers with brains, pronouncing that we understand, when the very process that produced us gives us no reason to believe that we have a basis for such understanding. Any tentative confidence we might have is derived from trial-and-error, sharing of stories, structured experiments, observations, mathematical models, and a continual challenging of each other, with, underneath, an unsubstantiated belief that the universe is not, at the very least, malignant or tricky. Excuse me, folks, but this is faith — a strong belief in things we hope for — by another word.

In the first half of the twentieth century, the scientist, paleontologist, geologist, and Jesuit priest Pierre Teilhard de Chardin wrote a book called *The Phenomenon of Man* (try to ignore the dated patriarchal language), with an introduction by British evolutionary biologist Sir Julian Huxley. To Darwin's naturalistic perspective of the physical world perceived through our bodily senses, Teilhard added a narrative of the emergence of internal complexity and personhood, which he based on evidence from paleontology and evolutionary biology. He was trying to come to grips with what we now call the mind, and the creativity of human societies. His interpretation of the data threatened both religious and scientific orthodoxy; the Catholic Church would not allow his books to be published in his lifetime, and the British defenders of scientific dogma, such as Peter Medawar, Steven Rose, and Richard Dawkins, called him a charlatan and purveyor of bad poetic science and deception.

In the introduction to *The Phenomenon of Man*, Huxley asserts that "we must infer the presence of a potential mind in all material systems, by backward extrapolation from the human phase to the biological." Emerging as it did within a tradition that thought it ridiculous to imagine that dogs were capable of suffering, or that elephants might show emotions, and that such things, although apparently observable, were clearly anthropomorphisms, this was a strong assertion. Now, of course, most reasonable scholars agree that what we see as the world is a function of both the observer and the observed, and that what we think of as consciousness, emotions, suffering, and culture among other animals are not mere anthropomorphisms. Indeed, recent investigations into rudimentary consciousness and the possibility of suffering in insect communities are consonant with Teilhard's earlier attempts to integrate the material and the experiential worlds.[123] Being a priest, he argued that "religion and science are the two conjugated faces or phases of the same complete act of knowledge."

The great twentieth-century philosopher and writer Arthur Koestler took a nonreligious stance on the subject, grounded in complexity and systems theories. In his books *The Ghost in the Machine* and *Janus: A Summing Up*, he asserted that anything we can think of — from atoms to arthropods to eco-social human societies — can be described in terms of Janus-faced holons. A holon is simultaneously a whole, composed of smaller elements, and a part of something larger. Seen as holons, we are individuals, made up of cells (which were probably originally single-celled life forms), and also members of eco-social systems that include plants, animals, soils, and social communities.

When Thomas Huxley's 1893 *Evolution and Ethics* was translated into Chinese, the Chinese characters used to render *evolution* (tian yan) could be read as "heaven's performance." What better way to speak of the unfolding universe? This is John Lennon's "Imagine" in other words. The evolutionary record gives evidence of the world's increasing complexity and our emergence within it, held together and bent by gravity, by strong and weak nuclear forces, and arriving at something like Regier's love and Wilson's biophilia. In Teilhard's words, "driven by the forces of love, the fragments of the world seek each other so that the world may come to being."

This way of thinking about the evidence offers a possible

reason *why* we should care, not just about science, but about the evolution of life on this planet. The fire that makes equations possible is both the alpha and the omega of the universe: the point at which evolution began and its ultimate end. More than that, since the fire is in us, and we are in the fire, "we are one, after all, you and I. Together we suffer, together exist, and forever will recreate each other." Since consciousness has been emerging as part of the evolutionary process, the fire is in the process of being created. Since this flame is in all things in the world, we all are participating in creating the world of the future.

This framing of our human conundrum is closer to Auden's vision, with which I started this chapter, than to John Lennon's beautiful but simple assertion that we should "imagine there's no heaven." This ecologically grounded understanding is not easily reconciled with the improbable and curmudgeonly Old Man painted by Michelangelo and rejected by Darwin, nor with the heaven rejected by Lennon. Michelangelo's image of the old patriarch is not the only one available, however — and it's probably the least scientifically and theologically interesting one. The fifth century BCE Atomists imagined a spatial model of a multiverse in which atoms scattered and regrouped into different formations. A couple of centuries later, the Stoics imagined a temporal universe, drying up and regrouping over time. In the fifteenth century CE, Nicholas of Cusa declared that the universe was without a center. Everything in the universe was in constant motion; because of this, no matter where you were, you yourself were at the center, with everything else moving around you. The universe, he argued, was almost infinite, just a little smaller than God, who was infinite. A century later, Giordano Bruno decided that, what the heck, the universe and the Creator were both infinite, which is a European version of Jainism. That little "what the heck" difference was noted by the church patriarchs. It was the difference that resulted in Cusa being made a cardinal and Bruno being burned at the stake. The narcissism of small differences between ideas, as much as between insect species, can have serious consequences. If these debates sound familiar, they are; except for the "God" part, the literature on modern physics is roiling with arguments over time, and space, and infinity, and how to reconcile them. As far as I can see, there is no definitive experiment or model in sight.

In the December 2015 Holiday Special edition of the *New Scientist*, Mary-Jane Rubenstein, a professor of religion at Wesleyan University, wrote that "if humans are not particularly godlike, then God is not particularly humanoid. God doesn't look like a patriarch in the sky: he looks like the universe." Rubenstein characterizes Cusa's and Bruno's forms of pantheism as "even more theologically threatening than atheism, precisely because they change what it means to be God. Not an anthropic creator beyond the world, but the force of creation within it." She adds that these ideas might cause us to "rethink what it is we mean by those godly terms like creation, power, renewal, and care. Is it possible that modern cosmology is asking us, not to abandon religion, but to think differently about what it is that gives life, what it is that's sacred, where it is we come from — and where we'll go?"

Entomophagy challenges us to ask the same questions. This is not just about a more sustainable food supply, or "learning" from insects, or acknowledging the services they provide. The many people who want us to "learn" from insects are very selective about which insects are allowed to be teachers: honey bees, for instance, if we are promoting cooperation, or ants, if we want to learn about hard work or engineering feats. Assassin bugs and Ichneumonidae? Not so much. If we consider the evolution of arthropods, and our emergence in the midst of an ecologically complex planet, then I am not sure how we can "learn" from nature. Usually what this learning means is that we are projecting our prior beliefs onto carefully selected species. We are in nature, and we are nature, and what we learn is no different from what we learn when we are mindful of the trillions of cells that live on, live in, and compose our own bodies. Some look at evolution and see competition among molecules, organisms, or groups. If we zoom in and out a few times using our imaginative, telescopic eyes, we see all of that; but, more stunningly, what we see is that we are integral parts of a world of molecules and organisms and landscapes that are evolving in ways that are multilayered and entangled. Because we are inside nature, the larger patterns and narratives are — like Mandelbrot's fractals[124] — embedded in us. We are created by, and continually re-create, the universe in which we dwell.

As Tim Flannery observed, the legacy in which we live is cooperative; nothing exists outside of relationships with other things. Where the nineteenth-century naturalists, rooted in the condescending, racist, patriarchal mentalities of the Empire, saw suffering and competition, a struggle to reproduce before dying, and an unfathomable or absent creative force, I see a world where every human life requires the taking of other lives, whether by eating them directly or by taking their food or claiming their habitats as our own. Where they saw no place for a fire, I find myself living in an emergent reality within which I have evolved, which I have helped, briefly, to shape, and to which I shall return. My obligation is to eat lightly, to cause as little harm as possible, but then also, in the end, like ants and termites and Ichneumonids, to return myself to the emergent biospheric community, so that others may eat, and live. Entomophagy is, for me, the naturalist's version of communion. It is a way of celebrating that what I am eating will someday eat me.

What I understand from the beetles — from all insects — and from their cooperative legacy, which dwells in us, in our DNA, and is the world in which we dwell, is that the force creating the once and future universe has no face. The creative force is not in the *individual* things (atoms, bacteria, plants, insects, mammals, people). The search for a particle where gravity dwells, or the organ that houses the mind (which René Descartes believed was the pineal gland), or the gut bacteria that influence our moods is, with regard to the question of the meaning of things, misplaced.

The force and the fire dwell in the dynamic, tense, unfolding relationships and conversations among us. The "force that through the green fuse drives the flower," in the inimitable words of Dylan Thomas, is comprised of trillions of rippling wavelets, dwells in the multichromatic appositional eye in which we are the lenses, and has many voices, discovering itself not by prescriptive pronouncement, but in conversation, speaking in tongues of magnetism and gravity to chemical molecules and wave-particles of light. The conversation looks like the aurora borealis, feels like biophilia, tastes like honey from the comb, asks of us that we eat and allows us the honor of in turn being eaten, so that the creating may continue, to an unknown end — which may perhaps not be a permanent ending but a contraction to a point and an explosion into a new universe.

I am happy with this understanding. In a world that often seems dark and fragmented, it gives us something to celebrate, and to ponder — a vision of who we could be, and why, with the power to motivate us to care about each other and this planet. The nuclear physicist Leo Szilard has said that an optimist is someone who believes that the future is uncertain. I am that kind of optimist. We have a voice in determining the nature of the world ahead of us. To me that is a huge motivation from day to day: building on that cooperative legacy to create a world that is like the heaven we wish for — or, in the words of Gandhi, to "be the change" we seek.

How, then, in our brief lives, can we nurture our best collective selves, as individuals within communities within the biosphere? Jeffrey Lockwood, reflecting on his research into the origins and disappearance of the devastating nineteenth-century locust plagues in America, said that for "the Rocky Mountain locust, the fertile river valleys of the mountainous West represented a sanctuary, a habitat where it could always find what it needed and persist in the face of adversity. We have such places too: churches, mosques, temples, and synagogues, along with hallowed groves, stone monoliths, and forested cathedrals. These sacred places comprise less than a millionth of the earth's surface but host three-quarters of the human population each year, and they are vital to our well-being."

I think about this when I consider our foraging and farming options for entomophagy. I am uncomfortable with mosques, churches, and synagogues as sanctuaries, since they too easily become militarized forts, full of fear. But that may be my hang-up, or maybe a general caution that any sanctuary, even an environmentally protected zone, can also become a base for a local militia in a misguided war. I am more at home in the oft-declared, and as often forgotten, belief that the creative fire is present everywhere, not in things or buildings, but in relationships among them. My sanctuary is looking out over a wide expanse of water, surrounded by little puff-clouds of bugs hovering just a few feet away, and the offshore breeze that prevents them from settling down and biting me.

I hope that in considering insects as food, we create and nurture and protect sanctuaries where we can respect and care for insects and ourselves together. I hope that we can have fun and enjoy adventurous eating with friends and companions, that we can find ways to fearlessly express our care for each other, for our enemies, and the planet that birthed us and that we are now co-creating.

# IN MY LIFE.
## ACKNOWLEDGMENTS

I am deeply indebted to the Canada Council for the Arts for helping to fund the research for this book through a Grant for Professional Writers, and also to the Ontario Arts Council for a Writers' Reserve Grant.

The chapter on ethics grew out of a long conversation I had with philosopher Karen Houle. Her book *Responsibility, Complexity, and Abortion: Toward a New Image of Ethical Thought* introduced me to the complexities of ethical thinking in this age of uncertainty and competing ideologies. So it was natural that I should turn to her for advice on how to deal with insects and entomophagy. Thanks, Dr. Houle, for helping me to find ways to get my head around this subject, for taking my naive questions seriously, challenging me to see the world in different, disturbing, and wonderful ways, and pushing all the appropriate brain buttons. In the spirit of Dr. Houle's ideas on complexity and ethics, I must say that, although the chapter would not have been written without her contributions, I take ownership of what I have written, including its misunderstandings and philosophically flawed arguments.

Special thanks to Christina Grammenos, my long-suffering and cheerfully diligent research assistant, who helped me track down many hundreds of books and papers, and then periodically summarized them for me when I was feeling swamped. Thanks to all the Facebook friends who riffed on Beatles' titles for me: Michael Bryson, Ainslie Butler, Dominique Charron, Dora Dueck, Shane Kurenoff, Judith Rosen, and especially that Beatlephiliac Massimo Rossetti. So many great suggestions. I'm sorry I couldn't use them all.

Thanks to Daniella Martin, Jeffrey Lockwood, Scott Shaw, and Alan Yen for responding to my questions, and for inspiring and informing me. All of you have written in such a way that I was able to dive off my retirement cliff into the entomological pool with great enthusiasm, knowing that I might hit my head on a rock if I dove shallow, drown if I dove too deep, and get the bends if I came up too fast. To Charlotte Payne, for translating emails from Japanese to English; for her passionate, groundbreaking, insightful, and diligent research; and for her assistance in exploring and understanding entomophagy worldwide, but especially in Japan. Thanks to Yukiko Kurioka of Japan Uni Agency for the arrangements she made to facilitate my explorations in Japan and to ensure I did not get lost. Thanks to the farmers, academics, and ento-entrepreneurs in Canada, France, Laos, Japan, and Australia who took the time to talk with me;

Jack David, Crissy Calhoun, and Laura Pastore at ECW for taking me on and challenging me to find the right words; and Kathy, for tolerating my warped enthusiasms.

Finally, I'd like to thank Mary Eleanor Bender, without whose prodding forty-five years ago I would never have tackled the topics in the final chapter. In the fall of 1970, at Goshen College, a small liberal arts college in Indiana, I would get up at eight in the morning to hear Mary Bender's lectures. The course was twentieth-century fiction. Tiny, single, elderly (I thought then, but probably in her fifties), she leaned over the podium and spoke quietly to us of Sartre, Camus, Ionesco, Kafka, Dos Passos, Woolf, Mansfield, Joyce, Robbe-Grillet — all those writers who helped define the twentieth-century European way of framing and grappling with the troubles of the world. I was riveted. And then, at the end of the course, she looked up from her podium at us — at *me* — and said, "They have defined the problem. Now it is up to you to find the solution." Her words became my life's vocation and passion. Professor Mary Eleanor, thank you. Sorry all this is a bit late to get a course credit.

## LETTING HER UNDER YOUR SKIN.
## RESTAURANTS, BUSINESSES, AND RECIPES

The entomophagy landscape continues to change rapidly and sometimes
unpredictably, so I list below just a few key websites and books that will
direct you to further sources on restaurants, businesses, and recipes.

### GENERAL SOURCES OF INFORMATION

This website is an excellent overall resource for current information:
http://www.scoop.it/t/entomophagy-edible-insects-and-the-future-
of-food

The documentary *Bugs on the Menu* gives an excellent overview,
and the follow-up Twitter feeds have some good ideas for recipes.
See http://bugsonthemenu.com/intro and https://twitter.com/
BugsontheMenu

## WHERE TO BUY BUGS FOR HUMAN CONSUMPTION

Daniella Martin's website

(https://edibug.wordpress.com/where-to-get-bugs/)

C-FU (These guys are new but look interesting!)

(http://cfufoods.com/#home)

Entomo Farms (http://entomofarms.com/)

Fédération Française des Producteurs Importateurs et Distributeurs
d'Insectes (http://www.ffpidi.org/)

## FEEDS FOR ANIMALS

General information

(http://4ento.com/2015/03/12/top-10-insect-feed-companies/)

Enterra Feed (Canada) (www.enterrafeed.com)

Ynsect (France) (http://www.ynsect.com/)

AgriProtein (South Africa) (http://www.agriprotein.com/)

Restaurants that serve insects seem to come and go like seasonal swarms. I have noted within the text several that I visited while doing research on this book and which seem to be navigating the fickle shoals of cultural food preferences, such as Vij's, Public, Billy Kwong, Archipelago. Rather than direct readers to phantoms or miss some really good new ones, I urge you to ask around your neighborhoods and cities and check out new venues on the web. The best bugs are probably close to home!

## WHERE TO LOOK. RECIPES

Recipes for insects in European and North American cuisine are an emerging phenomenon, and there are more being published every day. Some of these can be found in the complete bibliography on my website (www.davidwaltnertoews.wordpress.com).

Here are a few recent books with recipes for preparing insects:

Martin, Daniella. 2014. *Edible: An Adventure into the World of Eating Insects and the Last Great Hope to Save The Planet*. Boston: New Harvest, Houghton Mifflin Harcourt.

Nelson, Michelle. 2015. *The Urban Homesteading Cookbook: Forage, Farm, Ferment and Feast for a Better World*. Vancouver: Douglas & McIntyre.

van Huis, Arnold, Henk van Gurp, and Marcel Dicke. 2014. *The Insect Cookbook: Food for a Sustainable Planet*. Translated by Françoise Takken-Kaminker and Diane Blumenfeld-Schaap. New York: Columbia University Press.

There are also many sources on the web for insect-based recipes. Here are a few:

Bug Vivant (http://bugvivant.com/edible-insect-recipes/)

Cicada Invasion (http://cicadainvasion.blogspot.ca/2011/04/if-you-cant-beat-em-eat-em-cicada.html)

Entomo Farms (http://entomofarms.com/recipes/)

Girl Meets Bug (https://edibug.wordpress.com/recipes/)

Insects are Food (http://www.insectsarefood.com/recipes.html)

*The Telegraph* (http://www.telegraph.co.uk/foodanddrink/foodanddrinknews/10401191/Top-11-bug-recipes.html)

*Time* magazine (http://time.com/3830167/eating-bugs-insects-recipes/)

# BUG, BUG ME DO.
## SELECTED BIBLIOGRAPHY

My research for this book included reading through all or part of more than 600 scholarly and popular books, papers, and websites. You can find a more complete list of references on my website (www.davidwaltner toews.wordpress.com). The list below only includes sources from which I have taken direct quotes or which in my opinion are particularly noteworthy. They are in alphabetical order, by author's last name.

Arabena, Kerry-Ann. 2009. "Indigenous to the Universe: A Discourse on Indigeneity, Citizenship and Ecological Relationships" Thesis, Canberra: Australian National University. Available from: https://digitalcollections.anu.edu.au/handle/1885/9264

Bajželj, B., K.S. Richards, J.M. Allwood, P. Smith, J.S. Dennis, E. Curmi, and C.A. Gilligan. 2014. "Importance of Food-Demand

Management for Climate Mitigation." *Nature Climate Change*
4(10):924–929.

Barron, Andrew B., and Colin Klein. 2016. "What insects can tell us
about the origins of consciousness," *Proceedings of the National
Academy of Sciences* 113(18):4900–4808.

Belluco, Simone, Carmen Losasso, Michela Maggioletti, Michela,
Cristiana C. Alonzi, Maurizio G. Paoletti, and Antonia
Ricci. 2013. "Edible Insects in a Food Safety and Nutritional
Perspective: A Critical Review." *Comprehensive Reviews in Food
Science and Food Safety* 12(3):296–313.

Berenbaum, May Roberta. 1995. *Bugs in the System: Insects and Their
Impact on Human Affairs*. Reading, MA: Addison-Wesley.

Berenbaum, May Roberta. 2000. *Buzzwords: A Scientist Muses on Sex,
Bugs, and Rock 'n' Roll*. Washington, DC: Joseph Henry.

Berenbaum, May Roberta. 2009. "Insect Biodiversity – Millions and
Millions." Pp. 575–582 in *Insect Biodiversity: Science and Society*,
edited by R.G. Foottit and P.H. Adler. Hoboken, NJ: Wiley-
Blackwell.

Bodenheimer, Friederich Simon. 1951. *Insects as Human Food: A
Chapter of the Ecology of Man*. The Hague: W. Junk.

Brown, Valerie A., John A. Harris, and Jacqueline Y. Russell, eds.
2010. *Tackling Wicked Problems through the Transdisciplinary
Imagination*. London: Earthscan.

Brune, Andreas. 2014. "Symbiotic Digestion of Lignocellulose in
Termite Guts." *Nature Reviews Microbiology* 12(3):168–180.

Bukkens, Sandra F. 1997. "The Nutritional Value of Edible Insects."
*Ecology of Food and Nutrition* 36(2–4):287–319.

Cahill, Thomas. 1995. *How the Irish Saved Civilization: The Untold
Story of Ireland's Heroic Role from the Fall of Rome to the Rise of
Medieval Europe*. New York: Doubleday.

Campbell, Christy. 2006. *The Botanist and the Vintner: How Wine Was Saved for the World*. Chapel Hill, NC: Algonquin Books of Chapel Hill.

Cerritos, René, and Zenón Cano-Santana. 2008. "Harvesting Grasshoppers *Sphenarium purpurascens* in Mexico for Human Consumption: A Comparison with Insecticidal Control for Managing Pest Outbreaks." *Crop Protection* 27(3):473–480.

Cerritos Flores, R., R. Ponce-Reyes, and F. Rojas-García. 2015. "Exploiting a Pest Insect Species *Sphenarium purpurascens* for Human Consumption: Ecological, Social, and Economic Repercussions." *Journal of Insects as Food and Feed* 1(1):75–84.

Chen, Xiaoming, Ying Feng, and Zhiyong Chen. 2009. "Common Edible Insects and Their Utilization in China." *Entomological Research* 39:299–303.

Cifuentes-Ruiz, Paulina, Santiago Zaragoza-Caballero, Helga Ochoterena-Booth, Miguel Morón Rios. 2014. "A Preliminary Phylogenetic Analysis of the New World Helopini (Coleoptera, Tenebrionidae, Tenebrioninae) Indicates the Need for Profound Rearrangements of the Classification." *ZooKeys* 415:191–216.

Codex Alimentarius Commission. 2010. Development of Regional Standard for Edible Crickets and Their Products: Agenda Item 13, Seventeenth Session, Bali, Indonesia, November 22–26, 2010. Comments of Lao PDR. Food and Agriculture Organization of the United Nations and World Health Organization. Available from: ftp://ftp.fao.org/codex/Meetings/CCASIA/ccasia17/CRDS/AS17_CRD08x.pdf

Crittenden, Alyssa. 2011. "The Importance of Honey Consumption in Human Evolution." *Food and Foodways* 19(4):257–273.

Cruz-Rodríguez, J.A., E. González-Machorro, A.A. Villegas González, M.L. Rodríguez Ramírez, and F. Majía Lara. 2016.

"Autonomous Biological Control of *Dactylopius opuntiae*
(Hemiptera: Dactyliiopidae) in a Prickly Pear Plantation with
Ecological Management." *Environmental Entomology* 45(3):
642–648.

Dronamraju, K.R., ed. 1995. *Haldane's Daedalus Revisited*. Oxford:
Oxford University Press.

Dunn, David, and James P. Crutchfield. 2006. "Insects, Trees,
and Climate: The Bioacoustic Ecology of Deforestation and
Entomogenic Climate Change." Working Paper No. 2006-12-055.
Santa Fe Institute, Santa Fe, NM.

Durst, B., V. Johnson, R.N. Leslie, and K. Shono, eds. 2010.
*Forest Insects as Food: Humans Bite Back*. Bangkok: Food and
Agriculture Organization of the United Nations.

Durst, P.B., and Y. Hanboonsong. 2015. "Small-Scale Production of
Edible Insects for Enhanced Food Security and Rural Livelihoods:
Experience from Thailand and Lao People's Democratic Republic."
*Journal of Insects as Food and Feed* 1(1): 2531.

Erzinçlioglu, Zakaria. 2000. *Maggots, Murder and Men: Memories and
Reflections of a Forensic Entomologist*. Colchester, UK: Harley
Books.

European Food Safety Authority Scientific Committee. 2015. "Risk
Profile Related to Production and Consumption of Insects as
Food and Feed." *EFSA Journal* 13(10):4257. Available from:
http://www.efsa.europa.eu/en/efsajournal/pub/4257.

Evans, Edward P. 1906. *The Criminal Prosecution and Capital
Punishment of Animals*. London: William Heinemann. Available
from: http://www.gutenberg.org/files/43286/43286-
h/43286-h.htm

Feng, Y., and X. Chen. 2003. "Utilization and Perspective of Edible
Insects in China." *Forest Science and Technology* 44(4):19–20.

Flannery, Tim. 2010. *Here on Earth: A Natural History of the Planet*. Toronto: HarperCollins.

Gemeno, César, Giordana Baldo, Rachele Nieri, Joan Valls, Oscar Alomar, and Valerio Mazzoni. 2015. "Substrate-borne vibrational signals in mating communication of Macrolophus bugs." *Journal of Insect Behavior*. 28(4):482–498.

Glover, D., and A. Sexton. 2015. "Edible Insects and the Future of Food: A Foresight Scenario Exercise on Entomophagy and Global Food Security." Evidence Report No. 149, Institute of Development Studies. Available from: http://www.ids.ac.uk/publication/edible-insects-and-the-future-of-food-a-foresight-scenario-exercise-on-entomophagy-and-global-food-security

Golubkina, Nadezhda, Sergey Sheshnitsan, and Marina Kapitalchuk. 2014. "Ecological Importance of Insects in Selenium Biogenic Cycling." *International Journal of Ecology* 2014.

Gould, Stephen Jay. [1983] 1994. "Nonmoral Nature." Pp. 32–44 in *Hen's Teeth and Horse's Toes: Further Reflections in Natural History*. New York: W.W. Norton.

Goulson, Dave, Elizabeth Nicholls, Cristina Botías, and Ellen L. Rotheray. 2015. "Bee Declines Driven by Combined Stress from Parasites, Pesticides, and Lack of Flowers." *Science* 347(6229).

Halloran, Afton, Nanna Roos, and Yupa Hanboonsong. 2016. "Cricket Farming as a Livelihood Strategy in Thailand." *Geographic Journal*.

Halloran, A., P. Vantomme, Y. Hanboonsong, and S. Ekesi. 2015. "Regulating Edible Insects: The Challenge of Addressing Food Security, Nature Conservation, and the Erosion of Traditional Food Culture." *Food Security* 7(3):739–746.

Hanboonsong, Y., T. Jamjanya, and B. Durst. 2013. *Six-Legged Livestock: Edible Insect Farming, Collection and Marketing in*

*Thailand*. Bangkok: Food and Agriculture Organization of the United Nations. http://www.fao.org/docrep/017/i3246e/i3246e.pdf

Handley, M.A., C. Hall, E. Sanford, E. Diaz, E. Gonzalez-Mendez, K. Drace, R. Wilson, M. Villalobos, and M. Croughan. 2007. "Globalization, Binational Communities, and Imported Food Risks: Results of an Outbreak Investigation of Lead Poisoning in Monterey County, California." *American Journal of Public Health* 97(5):900–906.

Henry, M., L. Gasco, G. Piccolo, and E. Fountoulaki. 2015. "Review on the Use of Insects in the Diet of Farmed Fish: Past and Future." *Animal Feed Science and Technology* 203:1–22.

Houle, Karen. 2014. *Responsibility, Complexity, and Abortion: Toward a New Image of Ethical Thought*. Toronto: Lexington Books.

Hölldobler, Bert, and Edward O. Wilson. 2009. *The Super-Organism: The Beauty, Elegance, and Strangeness of Insect Societies*. New York: W.W. Norton.

Huang, H.T., and Pei Yang. 1987. "The Ancient Cultured Citrus Ant." *BioScience* 37(9):665–671.

Kanazawa, S., Y. Ishikawa, M. Takaoki, M. Yamashita, S. Nakayama, K. Kiguchi, R. Kok, H. Wada, and J. Mitsuhashi. 2008. "Entomophagy: A Key to Space Agriculture." *Advances in Space Research* 41(5):701–705.

Kinyuru, John N., Silvenus O. Konyole, Nanna Roos, Christine A. Onyango, Victor O. Owino, Bethwell O. Owuor, Benson B. Estambale, Henrik Friis, Jens Aagaard-Hansen, and Glaston M. Kenji. 2013. "Nutrient Composition of Four Species of Winged Termites Consumed in Western Kenya." *Journal of Food Composition and Analysis* 30(2):120–124.

Klunder, H.C., J. Wolkers-Rooijackers, J.M. Korpela, and M.J.R.

Nout. 2012. "Microbiological Aspects of Processing and Storage of Edible Insects." *Food Control* 26(2):628–631.

Lemelin, Rayland Harvey, ed. 2013. *The Management of Insects in Recreation and Tourism*. Cambridge: Cambridge University Press.

Lockwood, Jeffrey A. 1987. "The Moral Standing of Insects and the Ethics of Extinction." *Florida Entomologist* 70(1):70–89.

Lockwood, Jeffrey A. 2004. *Locust: The Devastating Rise and Mysterious Disappearance of the Insect that Shaped the American Frontier*. New York: Basic Books.

Lockwood, Jeffrey A. 2011. "The Ontology of Biological Groups: Do Grasshoppers Form Assemblages, Communities, Guilds, Populations, or Something Else?" *Psyche* 2011.

Lockwood, Jeffrey A. 2013. *The Infested Mind: Why Humans Fear, Loathe, and Love Insects*. Oxford: Oxford University Press.

Long, John A., Ross R. Large, Michael S.Y. Lee, Michael J. Benton, Leonid V. Danyushevsky, Luis M. Chiappe, Jacqueline A. Halpin, David Cantrill, and Bernd Lottermoser. 2015. "Severe Selenium Depletion in the Phanerozoic Oceans as a Factor in Three Global Mass Extinction Events." *Gondwana Research*.

Looy, Heather, Florence Dunkel, and John Wood. 2014. "How Then Shall We Eat? Insect-Eating Attitudes and Sustainable Foodways." *Agriculture and Human Values* 31(1):131–141.

Looy, Heather, and John R. Wood. 2015. "Imagination, Hospitality and Affection: The Unique Legacy of Food Insects?" *Animal Frontiers* 5(2):8–13.

Losey, John E., and Mace Vaughan. 2006. "The Economic Value of Ecological Services Provided by Insects." *BioScience* 56(4): 311–323.

Lundy, Mark E., and Michael Parrella. 2015. "Crickets Are Not a Free Lunch: Protein Capture from Scalable Organic Side-Streams via

High-Density Populations of *Acheta domesticus*." *PLoS ONE* 10(4):1–12.

Madsen, David B., and Dave N. Schmitt. 1998. "Mass Collecting and the Diet Breadth Model: A Great Basin Example." *Journal of Archaeological Science* 25(5):445–455.

Makhado, Rudzani, Martin Potgieter, Jonathan Timberlake, and Davison Gumbo. 2014. "A Review of the Significance of Mopane Products to Rural People's Livelihoods in Southern Africa." *Transactions of the Royal Society of South Africa* 69(2):117–122.

Martin, Daniella. 2014. *Edible: An Adventure into the World of Eating Insects and the Last Great Hope to Save the Planet*. Boston: New Harvest, Houghton Mifflin Harcourt.

McGrew, William C. 2014. "The 'Other Faunivory' Revisited: Insectivory in Human and Non-Human Primates and the Evolution of Human Diet." *Journal of Human Evolution* 71:4–11.

Nowak, Verena, Diedelinde Persijn, Doris Rittenschober, and U. Ruth Charrondiere. 2016. "Review of Food Composition Data for Edible Insects." *Food Chemistry* 193:39–46.

Oonincx, D.G.A.B., and I.J.M. de Boer. 2012. "Environmental Impact of the Production of Mealworms as a Protein Source for Humans — A Life Cycle Assessment." *PLoS ONE* 7(12): e51145.

Paoletti, Maurizio, Erika Buscardo, and Darna Dufour. 2000. "Edible Invertebrates among Amazonian Indians: A Critical Review of Disappearing Knowledge." *Environment, Development and Sustainability* 2(3):195–225.

Paoletti, Maurizio G., Lorenzo Norberto, Roberta Damini, and Salvatore Musumeci. 2007. "Human Gastric Juice Contains Chitinase That Can Degrade Chitin." *Annals of Nutrition and Metabolism* 51(3):244–251.

Payne, C.L.R. 2015. "Wild Harvesting Declines as Pesticides and

Imports Rise: The Collection and Consumption of Insects in Contemporary Rural Japan." *Journal of Insects as Food and Feed* 1(1):57–65.

Payne, C.L.R., P. Scarborough, M. Rayner, and K. Nonaka. 2015. "Are Edible Insects More or Less 'Healthy' than Commonly Consumed Meats? A Comparison Using Two Nutrient Profiling Models Developed to Combat Over- And Undernutrition." *European Journal of Clinical Nutrition.*

Payne, Charlotte L.R.; Peter Scarborough, Mike Rayner, and Kenichi Nonaka. 2016. "A Systematic Review of Nutrient Composition Data Available for Twelve Commercially Available Edible Insects, and Comparison with Reference Values." *Trends in Food Science & Technology* 47: 69–77.

Payne, Charlotte L.R., Mitsutoshi Umemura, Shadreck Dube, Asako Azuma, Chisato Takenaka, and Kenichi Nonaka. 2015. "The Mineral Composition of Five Insects as Sold for Human Consumption in Southern Africa." *African Journal of Biotechnology* 14(31): 2443–2448.

Pearson, Gwen. 2015. "You Know What Makes Great Food Coloring? Bugs." *Wired*, September 10. Available from: http://www.wired.com/2015/09/cochineal-bug-feature/

Pham, Hanh T., Max Bergoin, and Peter Tijssen. 2013. "*Acheta domesticus* Volvovirus, a Novel Single-Stranded Circular DNA Virus of the House Cricket." *Genome Announcements* 1(2): e0007913.

Plotnick, Roy, Jessica Theodor, and Thomas Holtz. 2015. "Jurassic Pork: What Could a Jewish Time Traveler Eat?" *Evolution: Education and Outreach* 8(17):1–14.

Popescu, Agatha. 2013. "Trends in World Silk Cocoons and Silk Production and Trade, 2007–2010." *Lucrari Stiintifice: Zootehnie si Biotehnologii* 46(2):418–423.

Premalatha, M., Tasneem Abbasi, Tabassum Abbasi, and S.A. Abbasi. 2011. "Energy-Efficient Food Production to Reduce Global Warming and Ecodegradation: The Use of Edible Insects." *Renewable and Sustainable Energy Reviews* 15:4357–4360.

Quammen, David. 2003. *Monster of God: The Man-Eating Predator in the Jungles of History and the Mind.* New York: W.W. Norton.

Raffles, Hugh. 2010. *Insectopedia.* New York: Patheon Books.

Rains, Glen C., Jeffery K. Tomberlin, and Don Kulasiri. 2008. "Using Insect Sniffing Devices for Detection." *Trends in Biotechnology* 26(6):288–294.

Ramos-Elorduy, Julieta. 2009. "Anthropo-Entomophagy: Cultures, Evolution and Sustainability." *Entomological Research* 39(5): 271–288.

Ramos-Elorduy Blasquez, Julieta, Jose Manuel Pino Moreno, Victor Hugo Martinez Camacho. 2012. "Could Grasshoppers Be a Nutritive Meal?" *Food and Nutrition Sciences* 3(2):164–175.

Raubenheimer, David, and Jessica M. Rothman. 2013. "Nutritional Ecology of Entomophagy in Humans and Other primates." *Annual Review of Entomology* 58:141–60.

Regier, Jerome C., Jeffrey W. Shultz, Andreas Zwick, April Hussey, Bernard Ball, Regina Wetzer, Joel W. Martin, and Clifford W. Cunningham. 2010. "Arthropod Relationships Revealed by Phylogenomic Analysis of Nuclear Protein-Coding Sequences." *Nature* 463:1079–1083.

Rinaudo, Marguerite. 2006. "Chitin and Chitosan: Properties and Applications." *Progress in Polymer Science* 31(7):603–632.

Rittell, Horst W.J., and Melvin Webber. 1973. "Dilemmas in a General Theory of Planning. *Policy Sciences* 4:155–169.

Roffet-Salque, Mélanie, Martine Regert, Richard P. Evershed, Alan K. Outram, Lucy J.E. Cramp, Orestes Decavallas, Julie Dunne, Pascale Gerbault, Simona Mileto, Sigrid Mirabaud, et al. 2015.

"Widespread Exploitation of the Honeybee by Early Neolithic Farmers." *Nature* 527:226–230.

Rothenberg, David. 2013. *Bug Music: How Insects Gave Us Rhythm and Noise*. New York: St. Martin's Press.

Sánchez-Muros, María-José, Fernando G. Barroso, and Francisco Manzano-Agugliaro. 2014. "Insect Meal as Renewable Source of Food for Animal Feeding: A Review." *Journal of Cleaner Production* 65:16–27.

Scientific Committee of the Federal Agency for the Safety of the Food Chain (SciCom) and the Board of the Superior Health Council. 2014. "Food Safety Aspects of Insects Intended for Human Consumption." Sci Com dossier 2014/04; SHC dossier n° 9160. Available from: http://www.health.belgium.be/en/food-safety-aspects-insects-intended-human-consumption-shc-9160-fasfc-sci-com-201404

Shaw, Scott Richard. 2014. *Planet of the Bugs: Evolution and the Rise of Insects*. Chicago: University of Chicago Press.

Shelomi, Matan. 2015. "Why We Still Don't Eat Insects: Assessing Entomophagy Promotion through a Diffusion of Innovations Framework." *Trends in Food Science & Technology* 45(2):311–318.

Skinner, Mark. 1991. "Bee Brood Consumption: An Alternative Explanation for Hypervitaminosis A in KNM-ER 1808 (*Homo erectus*) from Koobi Fora, Kenya." *Journal of Human Evolution* 20(6):493–503.

Smetana, S., A. Mathys, and V. Heinz. 2015. "Challenges of Life Cycle Assessment for Insect-Based Feed and Food." *INSECTA 2015 National Symposium on Insects for Food and Feed*. Available from: https://www.researchgate.net/publication/282085709_Challenges_of_Life_Cycle_Assessment_for_insect-based_feed_and_food

Strausfeld, Nicholas J., and Frank Hirth. 2013. "Deep Homology
of Arthropod Central Complex and Vertebrate Basal Ganglia."
*Science* 340(6129):157–161.

Szelei, J., J. Woodring, M.S. Goettel, G. Duke, F.-X. Jousset, K.Y.
Liu, Z. Zadori, Y. Li, E. Styer, D.G. Boucias, R.G. Kleespies, M.
Bergoin, and P. Tijssen. 2011. "Susceptibility of North-American
and European Crickets to *Acheta domesticus* Densovirus
(AdDNV) and Associated Epizootics." *Journal of Invertebrate
Pathology* 106(3):394–399.

Thomas, Benisiu. 2013. "Sustainable Harvesting and Trading
of Mopane Worms (*Imbrasia belina*) in Northern Namibia: An
Experience from the Uukwaluudhi Area." *International Journal of
Environmental Studies* 70(4):494–502.

Tomberlin, J.K., A. van Huis, M.E. Benbow, H. Jordan, D.A. Astuti,
D. Azzollini, I. Banks, V. Bava, C. Borgemeister, J.A. Cammack,
et al. 2015. "Protecting the Environment through Insect Farming
as a Means to Produce Protein for Use as Livestock, Poultry, and
Aquaculture Feed." *Journal of Insects as Food and Feed* 1(4): 307–309.

Tomotake, Hiroyuki, Mitsuaki Katagiri, and Masayuki Yamato. 2010.
"Silkworm Pupae (*Bombyx mori*) Are New Sources of High
Quality Protein and Lipid." *Journal of Nutritional Science and
Vitaminology* 56(6):446–448.

United States Food and Drug Administration. 1995. *Defect Levels
Handbook*. Available from: http://www.fda.gov/food/
guidanceregulation/guidancedocumentsregulatoryinformation/
ucm056174.htm

van Huis, A. 2014. "The global impact of insects." Farewell address
upon retiring as Professor of Tropical Entomology, Wageningen
University, November 20. Available from: http://www.academia
.edu/11840536/The_global_impact_of_insects

van Huis, Arnold, Henk van Gurp, and Marcel Dicke. 2014. *The Insect Cookbook: Food for a Sustainable Planet*. Translated by F. Takken-Kaminker and D. Blumenfeld-Schaap. New York: Columbia University Press.

van Huis, Arnold, Joost Van Itterbeeck, Harmke Klunder, Esther Mertens, Afton Halloran, Giulia Muir, and Paul Vantomme. 2013. *Edible Insects: Future Prospects for Food and Feed Security*. FAO Forestry Paper 171. Rome: Food and Agriculture Organization of the United Nations. Available from: http://www.fao.org/docrep/018/i3253e/i3253e00.htm

Waldbauer, Gilbert. 2003. *What Good Are Bugs? Insects in the Web of Life*. Cambridge, MA: Harvard University Press.

Waldbauer, Gilbert. 2009. *Fireflies, Honey, and Silk*. Los Angeles: University of California Press.

Waltner-Toews, David. 2004. *Ecosystem Sustainability and Health*. Cambridge: Cambridge University Press.

Waltner-Toews, David. 2007. *The Chickens Fight Back: Pandemic Panics and Deadly Diseases That Jump from Animals to Humans*. Vancouver, BC: Greystone.

Waltner-Toews, David. 2008. *Food, Sex, and Salmonella: Why Our Food Is Making Us Sick*. Vancouver, BC: Greystone.

Waltner-Toews, David, James J. Kay, and Nina-Marie E. Lister. 2008. *The Ecosystem Approach*. New York: Columbia University Press.

Webster, Timothy H., William C. McGrew, Linda F. Marchant, Charlotte L.R. Payne, and Kevin D. Hunt. 2014. "Selective Insectivory at Toro-Semliki, Uganda: Comparative Analyses Suggest No 'Savanna' Chimpanzee Pattern." *Journal of Human Evolution*. 71: 20–27.

Winston, Mark L. 2014. *Bee Time: Lessons from the Hive*. Cambridge, MA: Harvard University Press.

Wrightson, Kendall. 1999. "An Introduction to Acoustic Ecology." *Soundscape* 1:10–13.

Xu, Lijia, Huimin Pan, Qifang Lei, Wei Xiao, Yong Peng, and Peigen Xiao. 2013. "Insect Tea, a Wonderful Work in the Chinese Tea Culture." *Food Research International* 53:629–635.

Yates-Doerr, Emily. 2015. "The World in a Box? Food Security, Edible Insects, and 'One World, One Health' Collaboration." *Social Science & Medicine* 129:106–112.

Yen, A.L. 2012. "Edible Insects and Management of Country." *Ecological Management & Restoration* 13(1):97–99.

# ENDNOTES

NOTE: *Authors and dates referred to in these notes are in the Selected Bibliography.*

## CRICKET TO RIDE: AN INTRODUCTION

1.	van Huis et al. 2013. *Edible Insects: Future Prospects for Food and Feed Security*.

2.	The term *brood* refers to the young, which in the case of bees, hornets, and wasps means the larvae. Sometimes these are eaten together with the comb in which they are reared, and sometimes they are extracted from the comb and eaten separately.

3.	"Bugs in the System," *The Economist* (September 16, 2014). http://www.economist.com/news/science-and-technology/21620560-merits-and-challenges-turning-bugs-food-insect-mix-and-health

4.	Protein concentrates in feeds for chickens, farmed fish, and other livestock are often made up of "fishmeal," a euphemism for anchovies, pillaged from the ocean off the coast of Peru and then ground up in a messy industrial system. Soybeans, once enthusiastically proposed as an environmentally friendly alternative to meat, and then used as a protein concentrate in animal feeds, are now grown on cleared rainforest lands in Brazil.

5.	Until 2015, Entomo Farms was called Next Millennium Farms. This was their name when I first visited them. In order to avoid confusion, I shall refer to them as Entomo Farms throughout this book. The change of name reflects the changing fortunes of insect-eating in North American culture.

6. This definition is from Charles Doyle, *A Dictionary of Marketing,* 3rd ed. (Oxford: Oxford University Press, 2011). Christensen himself now prefers the term *disruptive innovation.* See http://www.claytonchristensen.com/key-concepts/

7. For spelling the common name of insects in this book, I have tried to follow the recommendations of the Entomological Society of America, which uses two words for honey bees, despite the common usage of running them together. The latter, they assert, is the equivalent of using Johnsmith, rather than John Smith. For more on this, see the database of common names of insects at http://entsoc.org/common-names

8. Haldane's original and oft-repeated quote is that life is "not only queerer than we suppose, but queerer than we *can* suppose," his use of the word queer for "strange" in this context being an illustration of the changing complexities of language and culture.

9. van Huis, 2014. "The Global Impact of Insects."

10. van Huis et al., 2013. *Edible Insects: Future Prospects for Food and Feed Security.*

## PART I. MEET THE BEETLES!

11. Yen, 2012. "Edible Insects and Management of Country."

12. Lockwood, 2011. "Ontology of Biological Groups."

13. To remember this classification system, they often use a mnemonic such as Dear King Phillip Came Over From Great Spain (or For Great Sex, or For Great Spaghetti, or one of those other S words).

14. Cifuentes-Ruiz et al., 2014. "Preliminary Phylogenetic Analysis."

15. For instance, an FDA handbook on foodborne pathogenic microorganisms and natural toxins is titled *The Bad Bug Book.* See http://www.fda.gov/Food/FoodborneIllnessContaminants/CausesOfIllnessBadBugBook/

16. See entomofarms.com

17. Bill Holm, *Boxelder Bug Variations: A Meditation on an Idea in Music and Language* (Minneapolis, MN: Milkweed Editions, 1985).

18. Durst et al., 2010. *Forest Insects as Food.*

19. Bukkens, 1992. "The Nutritional Value of Edible Insects."

20. Ramos-Elorduy et al., 2012. "Could Grasshoppers Be a Nutritive Meal?"

21. Payne et al., 2015. "Are Edible Insects More or Less 'Healthy' Than Commonly Consumed Meats?"

22. Cited in Durst et al., 2010. *Forest Insects as Food.*

23. Nowak et al., 2016. "Review of Food Composition Data for Edible Insects."

24. In South Africa, de-gutting mopane worms has been described as being like squeezing the contents out of a tube of toothpaste.

25. Payne et al, "Nutrient Composition Data."

26. "How Eating Insects Could Help Climate Change," *BBC News* (December 11, 2015). http://www.bbc.com/news/science-environment-35061609

27. Paoletti et al, 2007.

28. Although they look like beetles, they are, like all cockroaches, in the order Blattodea, along with termites.

29. Taking this work a step further, in 2016 Natasha Grimard, a seventeen-year-old Canadian student, proposed that insect-fortified foods could be used to enhance the nutritional status of people in refugee camps, and created foods that would be culturally appropriate. The following year, she received a prestigious national award for innovation. See https://www.youtube.com/watch?v= ZCCytkR-YqE

30. "How Eating Insects Empowers Women," *Motherboard* (April 18, 2016). http://motherboard.vice.com/read/how-eating-insects-empowers-women

31. Lundy and Parella, "Not a Free Lunch."

32. Reports comparing FCR for crickets to other species are strongly contested by insect farmers. My point is that the advantages are not always as clear as one might think.

33. Bajželj et al, "Importance of Food-Demand Management."

34. See *Livestock's Long Shadow*, for instance, an FAO report published in 2006 and available online.

35. I mentioned this interview earlier, when discussing chitinase (http://www.bbc .com/news/science-environment-35061609).

36. For a fuller discussion of some of the challenges, strategies, and options, see David Waltner-Toews, *Ecosystem Sustainability and Health* (Cambridge: Cambridge University Press, 2004) and David Waltner-Toews, James J. Kay, and Nina-Marie E. Lister, *The Ecosystem Approach* (New York: Columbia University Press, 2008).

37. J. Fernquest, "Eating Insects: Sudden Popularity," *Bangkok Post* (May 31, 2013). http://www.bangkokpost.com/learning/learning-news/352836/eating-insects-sudden-popularity

## PART II. YESTERDAY AND TODAY: INSECTS AND THE ORIGINS OF THE MODERN WORLD

38. The idea that the earth's tectonic plates moved was first proposed by Alfred Wegener early in the twentieth century. Wegener was ridiculed by fellow scientists for suggesting this — until he was proven correct in the 1950s, after he had died.

39. In this insectoid tradition, I composed a song and performed it at a folk festival in 1970. The lovely young woman for whom it was written subsequently agreed to marry me. I know it's association, not causation, but still . . .

40. Webster et al., 2014. "Selective Insectivory at Toro-Semliki Uganda."

41. W. Bostwick, "Boiled Alive: Turning Bees into Mead," *Food Republic*

(September 20, 2011). http://www.foodrepublic.com/2011/09/20/boiled-alive-turning-bees-into-mead/

42.   In this book, I have used *tons* to refer to both the American weight (2,000 pounds) and the international metric weight (1,000 kilograms or 2,200 pounds). Because the production figures on insects and insect products are estimates, and most are increasing, this difference does not alter the general arguments I am making.

43.   Crittenden, 2011. "The Importance of Honey Consumption in Human Evolution."

44.   Flannery, 2010. *Here on Earth: A Natural History of the Planet.*

45.   Latin for foolish, but then, you knew that already.

46.   Difficulty in degrading lignocellulose is one of the big challenges facing the twenty-first-century biofuel industry.

47.   See http://michaelpollan.com/reviews/how-to-eat/

48.   Berenbaum, 1994. *Bugs in the System.*

49.   See www.unspunhoney.com.au

50.   Winston, 2014. *Bee Time*, p222.

51.   Rothenberg, 2013. *Bug Music: How Insects Gave Us Rhythm and Noise*, p8.

52.   Dunn and Crutchfield, 2006. "Insects, Trees, and Climate."

53.   Raffles, 2010. *Insectopedia*, p316.

## PART III. I ONCE HAD A BUG: HOW PEOPLE CREATED INSECTS

54.   This is evident even when the cliche is turned on its head and turned into self-mocking horror, such as in movies like *Meet the Applegates*. The humor only works because of the initial gut reaction.

55.   Quammen, 2003. *Monster of God*, p431.

56.   The World Organisation for Animal Health is referred to by those in the know as the OIE, a historic hangover from 1924, when this organization was founded in Paris as the Office International des Epizooties.

57.   W. Grimes, "When Bugs Declared Total War on Wine," *New York Times* (March 26, 2005). http://www.nytimes.com/2005/03/26/books/when-bugs-declared-total-war-on-wine.html?_r=1

58.   M. Gladwell, "The Mosquito Killer," *Gladwell.com* (July 2, 2001). http://gladwell.com/the-mosquito-killer/

59.   It's sometimes implied that "pulling the plug" on someone on a respirator at a hospital is "playing God." In fact, as Paneloux would frame it, it's when we put in the plug that we are "playing God."

60.   Readers interested in more titillating details may wish to consult Edward, *Criminal Prosecution and Capital Punishment of Animals.*

61.   I dealt with some of these complexities with regard to insect-borne zoonoses in my book *The Chickens Fight Back* (Vancouver, BC: Greystone, 2007).

62.   For more on heptachlor and aldicarb in the food chain, see my book *Food, Sex and Salmonella: Why Our Food Is Making Us Sick* (Vancouver, BC: Greystone, 2008).

63.    "Bayer Agrees to Terminate All Uses of Aldicarb," United States
       Environmental Protection Agency (August 17, 2010). https://yosemite.epa.
       gov/opa/admpress.nsf/e51aa292bac25b0b85257359003d925f/29f9dddede
       97caa88525778200590c93!OpenDocument

64.    See http://www.aglogicchemical.com/about.html

65.    A. Vowels, "Plan Bee," *The Portico* (September, 2015). https://www.uoguelph.ca/
       theportico/archive/2015/PorticoSum2015.pdf

## PART IV. BLACK FLY SINGING: REIMAGINING INSECTS

66.    Hölldobler and Wilson, 2009. *Superorganism: The Beauty, Elegance and
       Strangeness of Insect Societies*, p486.

67.    "The Composer and Conductor Mr. Fung Liao," *Bolingo.org* (July 28, 2006).
       http://bolingo.org/cricket/mrfung.htm

68.    *Cicada Invasion Survival Guide*. http://cicadainvasion.blogspot.ca/

69.    L. Bridget, "Fleas Are for Lovers," *Ploughshares* (June 9, 2010).
       http://blog.pshares.org/index.php/fleas-are-for-lovers/

70.    Raffles, 2010. *Insectopedia*, p343.

71.    In fact, the last known swarm of *Patanga succincta* was in 1908, in India. Its
       pestiferous activities in Thailand were thus more "grasshopper-like."

72.    Cerritos and Cano-Santana, 2008. "Harvesting Grasshoppers *Sphenarium
       purpurascens* in Mexico for Human Consumption."

73.    Cruz-Rodríguez et al., 2016. "Autonomous Biological Control."

74.    They are classified as Coccinellidae, a family of beetles belonging to the sub-
       order Polyphaga.

75.    Lockwood, 2013. *Infested Mind*, p171.

## PART V. GOT TO GET YOU INTO MY LIFE

76.    Xu et al., 2013. "Insect Tea."

77.    Cahill, 1995. *How the Irish Saved Civilization*, p217.

78.    Rittell and Webber, 1973. "Dilemmas in a General Theory of Planning." See also
       Brown et al., 2010. *Tackling Wicked Problems*.

79.    Flannery, 2010. *Here on Earth*, pp126–127.

80.    The name *SLA* was drawn from "symbiosis," which SLA leader Donald
       DeFreeze defined as "a body of dissimilar bodies and organisms living in deep
       and loving harmony and partnership in the best interest of all within the body."
       At least in the abstract sense, then, the two SLAs appear to be similar.

81.    See http://www.aspirefg.com/

82.    C. Matthews, "Bugs on the Menu in Ghana as Palm Weevil Protein Hits the
       Pan. *The Guardian* (January 3, 2016). https://www.theguardian.com/global-
       development/2016/jan/03/bugs-eat-insects-palm-weevils-ghana-protein

83.    Kinyuru et al., 2013. "Nutrient Composition of Four Species of Winged
       Termites."

84. The Japanese will sometimes add the suffix *-san* after a name, which is a gender-neutral honorific with no real English counterpart, perhaps comparable to a *Mr.* or *Ms.*, but more respectful. In this book, I have used the name and form of address most used by the people I talked to and their friends. Yukiko, for instance, having been educated in Canada and worked in the international book trade, asked that I simply call her by her first name.

85. As a bonus, I got to see kaffir limes, whose leaves I had used for cooking at home; I had wondered what the fruits looked like, having never seen them before. They look like wrinkly limes, a little like small guavas, and are apparently used to make soap and shampoo and other personal cleaning products — as well as for cooking stir-fries and curries.

86. LAK = Lao kip, the local currency; 1,000 LAK is about 12 US cents. So if he had eight cycles with ten kilograms per cycle, he would pull in just over US$300/year. Not a lot, but in Lao PDR, a living.

87. See http://www.theorganicprepper.ca/updated-prepping-for-an-ebola-lockdown-10022014

88. Tomberlin et al., 2015. "Protecting the Environment through Insect Farming."

89. See http://www.enterrafeed.com/about/#history. Although not mentioned on the website, soybeans, once proclaimed as the great alternative to livestock meat that was going to save the planet, are now an important driver of deforestation in South America.

90. On July 20, 2016, Enterra announced that, after four years of testing and review, CFIA had officially approved their product as an ingredient for chicken feed.

91. F. Tarannum, "Crickets the New Chicken? That's Chef Meeru Dhalwala's Mission," *The Tyee* (July 30, 2015). http://thetyee.ca/Culture/2015/07/30/Edible-Crickets/

92. The Russian Mennonite name for perogies.

93. S. Killingsworth, "Tables for Two: The Black Ant," *The New Yorker* (August 24, 2015). http://www.newyorker.com/magazine/2015/08/24/tables-for-two-the-black-ant

94. J. De Graaff, "Dishing Up Insects with Kylie Kwong," *Broadsheet* (May 1, 2013). http://www.broadsheet.com.au/melbourne/food-and-drink/article/dishing-insects-kylie-kwong-billy

95. J. Branco, "Entomophagy, a Pint of Science and the Men Who Want You to Eat Bugs." *Brisbane Times* (May 20, 2015). http://www.brisbanetimes.com.au/queensland/entomophagy-a-pint-of-science-and-the-men-who-want-you-to-eat-bugs-20150520-gh606x.html

## PART VI. REVOLUTION 1

96. Schweitzer's "Reverence for Life," for which he received a Nobel Peace Prize, is similar to the beliefs of the Jains in India.

97. Michael Ignatieff, *The Needs of Strangers* (Toronto: Penguin, 1984), 13.

98. Henry Regier, "Ecosystem Integrity in a Context of Ecostudies as Related to the Great Lakes Region," in *Perspectives on Ecological Integrity*, eds. L. Westra and J. Lemon (Dordrecht: Kluwer, 1995), 88–101.

99. Lockwood, 1987. "The Moral Standing of Insects."

100. Barron et al., 2016. "What Insects Can Tell Us About the Origins of Consciousness."

101. "The Green Brain Project," http://greenbrain.group.shef.ac.uk/

102. K. Segedin, "The Unexpected Beauty of Bugs," *BBC* (May 1, 2015). http://www.bbc.com/earth/story/20150425-the-beautiful-bugs-of-earth-capture

103. "The Beautiful Bugs of Belize," *BBC* (March 16, 2015). http://www.bbc.com/earth/story/20150309-the-beautiful-bugs-of-belize

104. This resonates with Mark Winston's comment: "What is unique about our relationship with honey bees is not only how much we depend on their services but also the fact that their health and survival depend on how well we manage the environment on which they rely. If we had a formal contract with honey bees, its executive summary might read something like this: We, the bees, will provide you with honey and other products of the hive, as well as pollination services. In return you, the humans, will maintain an environment in which we can thrive, free of toxic pesticides and rich in diverse flowering plants." See Winston, *Bee Time*.

105. Rains et al., 2008. "Using Insect Sniffing Devices for Detection." There are also other papers on the subject.

106. Obviously not a study designed by an epidemiologist.

107. *Defect Levels Handbook*, United States Food and Drug Administration (2016). http://www.fda.gov/Food/GuidanceRegulation/GuidanceDocuments RegulatoryInformation/SanitationTransportation/ucm056174.htm

108. "Risk Profile Related to Production and Consumption of Insects as Food and Feed," *EFSA Journal* 13(10, 2016)): 4257, doi:10.2903/j.efsa.2015.4257. See https://www.efsa.europa.eu/en/efsajournal/pub/4257

109. O. Rousseau, "Industry Questions EU Insect Meat Food Law," *Meat Processing* (November 19, 2015). http://www.globalmeatnews.com/Safety-Legislation/ Industry-questions-EU-insect-meat-food-law

110. The World Bank stopped using the term in 2016.

111. This is true globally for most food imports, not just insects.

112. Waldbauer, 2009. *Fireflies, Honey, and Silk*, p37.

113. "A Point of View: On Bees and Beings," *BBC News* (June 3, 2012). http://www.bbc.com/news/magazine-18279345

114. See J.A. VanLeeuwen, D. Waltner-Toews, T. Abernathy, and B. Smit, "Evolving Models of Human Health Toward an Ecosystem Context," *Ecosystem Health* 5(3, 1999): 204–219. The figure is also discussed in my book *Ecosystem Sustainability and Health: A Practical Approach*.

115. Yates-Doerr, 2015. "The World in a Box?"

116. See http://www.archipelago-restaurant.co.uk

117. In September 2015, Redzepi announced that Noma would close at the end of 2016 and resurrect itself as an urban farm, following the seasonal patterns more closely. J. Gordinier, "René Redzepi Plans to Close Noma and Reopen It as an Urban Farm," *New York Times* (September 14, 2015). http://www.nytimes .com/2015/09/16/dining/noma-rene-redzepi-urban-farm.html?_r=0

118. Kanazawa et al., 2008. "Entomophagy: A Key to Space Agriculture." Insects don't take up much room and don't compete for human foods; they can be used to recycle waste, and can even, like silkworms, provide materials for clothing. With the right bugs, Andy Weir's Martian Mark Watney could have lasted a lot longer. Or maybe the crew were eating insects. I don't recall the astronauts' diets being made explicit.

119. Arabena, 2009. "Indigenous to the Universe."

## PART VII. REVOLUTION 9

120. Dronamraju, 1995. *Haldane's Daedalus Revisited.*

121. I am not sure if Hawking realized how his framing of this echoed that of Haldane.

122. Gould, 1983/1994. "Nonmoral Nature."

123. Stephen Jay Gould, a generation after Teilhard, argued for a separation of these two kinds of questions: how and why. He asserted that there were questions appropriate to the magisterium of science and questions appropriate to the magisterium of religion, and that these were separate "non-overlapping" magisteria. This still leaves open the question of how we go about integrating insights from these magesteria.

124. For more on fractals, see Benout Mandelbrot, *The Fractal Geometry of Nature* (New York: Times Books, 1982).

# YOU MIGHT SEE ME.
## INDEX

regulations

    re honey bees, 268

    re insect consumption, 257–272

    re insects-as-food (EU), 217–219, 263, 271–272

    reasons for, 266–267

    as reflecting social ideals, 237

Renewable Food for Animals and Plants™, 211

revulsion

    defined, 119

    roots of, 107

rice grasshoppers, 31

Riley, Charles Valentine, 18, 117, 162–163

river blindness, 121, 139

roaches, 54. *See also* cockroaches

rock art

    grasshoppers in, 63

    hive-harvesting in, 67

Rothenberg, David, 96, 144, 166

r-strategists, 20

## S

sago worms. *See* palm weevil larvae

*sal de hormiga*, 225–226

San people, 73

sanctuaries, 280–281, 312–313. *See also* insect sanctuaries

sawflies, 21, 56

scale insects

    California red scale (*Aonidiella aurantii*), 13

    cochineal scale insects, 221, 279

    cottony cushion scale (*Icerya purchasi*), 162–163

    exudate of, 131

scaling up

    challenges in, 203

    technology and, 233

scientists

    classification of insects by, 4–6, 8, 11

    on insect population size, 18–19

    place-based science and, 286–288

    preconceptions of, 123–124

    as skeptical of anecdotal claims, 27–28

scorpionflies, 55

screw-worm flies (*Cochliomyia hominivorax*), 164

selenium, 77–78

sericulture, 280. *See also* silkworms

sexual cannibalism, 122

Shakespeare, William, on flies, 108

Shaw, Scott Richard, 10

    on insect ears, 54

    on insect legs, 24

    on primates and insects, 63

sheep liver fluke (*Dicroceolium dendriticum*), 80

Shelomi, Matan, 200, 232–233

silkworm larvae/pupae

    allergic reactions to, 258

    with cassava, 259

    consumption of in Asia, 66, 183

    importing of, 278

    on the menu, 188

    nutrient levels in, 27, 32–33

silkworms

    consumption of, 188

    domestication of, 146, 279–282

    farming of, 25

    Feed Conversion Ratio (FCR) of, 183

    in literature, 140

    white mulberry and, 183

Siphonaptera, 9. *See also* fleas

*Six-Legged Livestock* (FAO 2013), 160

sleeping sickness, 111–113

social capital, 157

social Darwinism, 173

social insects

    ancestors of, 56

    as capable of suffering, 241

    eusocial insects, 25

human relationship with, 71–72, 142

language describing, 173

primates eating, 61–62

Soper, Fred, 125–126, 129

*The Sound of Light in the Trees: The Acoustic Ecology of Pinyon Pines,* 96–97

South America, 31, 38, 44, 63–64, 87, 89–90, 178, 276

Southeast Asia, 4, 31–32, 194

    Cambodia, 108–110, 212, 278

    Lao PDR, 94, 109–110, 187, 195–201, 278

    Thailand, 109–110, 153–154, 160, 178–179, 194–195, 229–230, 277–278

southern pine beetles, 165–166

*Souvenirs entomologiques,* 92

stag beetles, 248

Starbucks, cochineal insects incident, 221–222, 238

sterile insects in pest management, 163–164

Stewart, Amy, 121–123

stick insects, 24

stink bugs. *See also* true bugs

    citrus stink bugs, 159

    consumption of, 195

    as garnish, 229

    as invasive, 121

    preparation method for, 177, 259

storage and preservation of insects, 155–156, 201, 260, 274

suffering

    in animals, 240–241, 306–307

    in insects, 240–245, 249–250, 307

*The Superorganism: The Beauty, Elegance and Strangeness of Insect Societies,* 142, 249

superposition eyes, 98–99

superworms, 45–46

sushi, as once marginal food, 182–183

sustainability

    of agri-food, 42, 204, 207

    entomophagy and, 300

    of fisheries, 210–211

    of food security, 40, 48, 58, 128, 171, 219

    supply-side, *xviii,* 202

sustainable livelihoods, 176

Suzuki, David, 211, 215

swarm/swarming, 6, 106

    of bees, 93, 100

    in literature, 106, 150

    as moral problems, 127–128

    phases of, 85, 113–114

    size of, 20

Swift, Jonathan, on fleas, 20–21

symphylans, 52–53

synapsids, 55

## T

Tachinid flies, 81

taxonomies, 11, 237

temple wall paintings, beehives in, 68

Tenebrionidae, 13, 86

termite mound clay, 181

termite mushroom (*Termitomyces*), 79

termites

    decreasing habitat for, 43–44

    as detritivores, 78

    egg-laying habits of, 20

    as eusocial insects, 25

    fishing for, 62

    fungi farming by, 79–80

    higher termites, 79

    hominids eating, 62–63

    as keystone species, 78

    lower termites, 79

    *Macrotermes* spp, 181